Back to Basics

Woodworking

Machines

WITHDRAWN

Back to Basics

Woodworking Machines

Straight Talk for Today's Woodworker

© 2010 by Skills Institute Press LLC
"Back to Basics" series trademark of Skills Institute Press
Published and distributed in North America by Fox Chapel Publishing Company, Inc.

Woodworking Machines is an original work, first published in 2010.

Portions of text and art previously published by and reproduced under license with Direct Holdings Americas Inc.

ISBN 978-1-56523-465-9

Library of Congress Cataloging-in-Publication Data

Woodworking machines.
      p. cm. -- (Back to basics)
Includes index.
ISBN 978-1-56523-465-9
1. Woodworking tools. 2. Woodworking machinery. I. Fox Chapel Publishing.
TT186.W6582 2010
684'.082--dc22

                              2010003813

To learn more about the other great books from Fox Chapel Publishing, or to find a retailer near you, call toll-free 800-457-9112 or visit us at *www.FoxChapelPublishing.com*.

Note to Authors: We are always looking for talented authors to write new books in our area of woodworking, design, and related crafts. Please send a brief letter describing your idea to Acquisition Editor, 1970 Broad Street, East Petersburg, PA 17520.

Printed in China
First printing: May 2010

Because working with wood and other materials inherently includes the risk of injury and damage, this book cannot guarantee that creating the projects in this book is safe for everyone. For this reason, this book is sold without warranties or guarantees of any kind, expressed or implied, and the publisher and the author disclaim any liability for any injuries, losses, or damages caused in any way by the content of this book or the reader's use of the tools needed to complete the projects presented here. The publisher and the author urge all woodworkers to thoroughly review each project and to understand the use of all tools before beginning any project.

Contents

Introduction..8

Chapter 1: **Maintaining Stationary Tools**......................12

Chapter 2: **Portable Power Tools**............................16

Chapter 3: **Sharpening Blades and Bits**.......................30

Chapter 4: **Table Saw**.....................................56

Chapter 5: **Band Saw**.....................................94

Chapter 6: **Radial Arm Saw**...............................122

Chapter 7: **Drill Press**...................................148

Chapter 8: **Jointer**......................................168

Chapter 9: **Other Machines**...............................179

Index..188

What You Can Learn

Maintaining Stationary Tools, p. 12

Learning how to adjust your stationary machines properly will improve the results and increase your pleasure from them.

Portable Power Tools, p. 16

Whatever their price range or list of features, all portable power tools will work better and last longer if they are cared for properly.

Sharpening Blades and Bits, p. 30

In addition to cutting and shaping properly, well-sharpened blades and bits offer other benefits, including reduced wear and tear on motors, less operator fatigue, and longer life for the blades and bits themselves.

Table Saw, p. 56

The table saw is an excellent all-around tool for cutting wood to width (ripping) and length (crosscutting) but it also accepts a wide variety of blades and accessories, from roller stands that assist with unwieldy panels to molding heads capable of producing elaborate decorative trim.

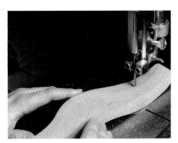

Band Saw, p. 94

In addition to crosscutting and ripping, the band saw is well suited for cutting curves and circles, enabling the woodworker to produce anything from a dovetail joint to a cabriole leg.

Radial Arm Saw, p. 122

With the exception of crosscutting very wide boards, radial arm saws can duplicate just about any job a table saw can perform but requires less workshop space and allows the piece to remain stationary.

Drill Press, p. 148

Originally designed for the metalworking trades, the drill press has found a second home in woodworking shops, where it enables woodworkers to bore precise holes and does duty as a sander and mortiser.

Jointer, p. 168

Any woodworker dedicated to precision and craftsmanship will attest that using this surfacing tool properly is the first step in turning rough boards into well-built pieces of furniture.

Other Tools, p. 179

Scroll saws, sanders, and air pumps are important tools for a woodworkers repertoir.

Table Saws

The first table saw I ever encountered belonged to my friend's dad, a Danish boat builder. He'd built it himself out of an assortment of parts of indeterminate age and origin and it proudly occupied a corner of his workshop. It always amazed me how Sven could turn out any number of identical, precisely cut pieces and then, after changing his set-up, make lengths of beautiful custom moldings. All these operations were done with seemingly effortless efficiency.

I was always pestering him to let me try using it and, when he finally did, the first project we made together was a set of floorboards for my vintage car, a 1938 Rover Sports Sedan. Even now, nearly 25 years later, I still look back at Sven and the floorboard project as the first stirrings of my desire to make a living from woodworking.

My present table saw, bought second-hand more than 10 years ago, is a 12-inch commercial model with a 3-horsepower motor. I couldn't imagine my woodworking shop without it. I rely on my table saw at many stages throughout my projects, whether cutting workpieces to size, making different joints, building drawers and doors, or creating a variety of molding patterns. I also make a lot of chairs in my workshop and I find my table saw particularly useful for cutting the angled tenons on the seat rails. When I built my house, I started by building the workshop; once that was up, the table saw came through for me once again, cutting sheets of plywood to size and making all the trim for the entire house, as well as cutting other assorted pieces.

I think I like the table saw so much for two main reasons. First of all, it's such a versatile machine; second, since most of the motor and blade are beneath the table, you can see at a glance what's going on. There's nothing to obscure your view of the work surface. Still, I have a great deal of respect for its ability to cause bodily injury—a lesson that's been drummed into me on a couple of occasions. But I consider it a safe machine, as long as the proper precautions are observed and the operator isn't overtired or in too much of a hurry. All in all, the table saw is a magnificent machine and I couldn't do without it.

- Giles Miller

A native of New Zealand, Giles Miller–Mead is seen here in his workshop with one of his prized tools—a vintage table saw acquired in the early 1980s.

Woodworking Machines

Drill Presses

Of all the tools in my shop, the drill press may not see as many hours of use as some others, but for certain tasks it is indispensable. The machine I use is a Sears Craftsman, manufactured in the mid-1950s. It came into our shop about four years ago. What I like about this drill press is its old-tool charm, the weight of it. It's solid. It was built to last.

Prior to getting into furniture-making, I experimented with carpentry and cabinetmaking. I studied woodworking in Colorado, taking classes with such highly respected furniture makers as Art Carpenter. Furniture-making is what I've been doing for the last seven years. I find it very satisfying. I'm constantly learning new techniques and trying new designs.

I produce a line of furniture as well as custom design pieces. I make a rocking chair for children with a design that includes bear paws on the arms and dowels set into a curved frame that provides back support. I depend on the drill press to bore precise holes for the dowels.

On a rocking chair I make for adults, the legs are square at the middle where they meet the seat, and have tenons at either end that fit into the rockers at the bottom and the arms at the top. To make the transition between the leg's square middle and round tenons, I sculpt the legs with a router and a spokeshave. The drill press reams the holes in the rockers and the arms for the tenons.

- *Judith Ames*

Judith Ames is a furniture-maker in Seattle, Washington.

Band Saws

I have been a Windsor chairmaker for 10 years. The machines in my shop are a lathe and a band saw. The lathe is essential; the bandsaw is a wonderful convenience. It cuts out seats, trims rough wood, cuts turnings and spindles to length, saws wedge slots, and is just plain handy. All these jobs could be done with hand saws, but the band saw does them quicker and more accurately.

I was exposed to band saws at an early age in my father's display and exhibit shop. Later I worked in a boatyard where the band saw made many wonderful shapes. That must have been where I realized it is my favorite woodworking machine. For a small shop limited to one stationary power tool, most folks would want a table saw, but a band saw would be my choice.

Obviously, band saws are great for cutting curves, down to tiny radii with $1/16$-inch blades. Angle cuts, straight or curved, are easy. With a little forethought, you can make "release cuts" and get into really tight places. Band saws also do a reasonable job of ripping with a $1/2$-inch or $3/4$-inch blade and a fence. My 14-inch Delta model will resaw boards 6 inches wide—any thickness from veneer on up. With a 6-inch riser block in its frame it could resaw 12 inches. That would take a monster industrial 30-inch table saw to make the same cut—or two passes with a 16-inch saw.

- Dave Sawyer

Dave Sawyer builds Windsor chairs at his workshop in South Woodbury, Vermont.

Jointers

The first jointer I used was already old when my father bought it. The machine required constant care, which was good, because it taught me to concentrate and pay attention to every board I worked with. When I was 15 years old, we built a new house and used locally grown oak for the trim and cabinets. It was my job to do all of the jointing.

A rule I learned then—and one that I still follow today—is that the success of any cabinetmaking project hinges on working with wood that has straight and square edges. The power tools that do most of the cutting in my shop nowadays—the table saw and the radial arm saw—will cut accurately only if the stock I feed into them is square and true. If one edge of a board is not straight, I won't be able to crosscut it squarely.

Getting off to a good start is where the jointer comes in. I use it to make that critically important first step, forming a square corner where the edge and the end of a board meet. The jointer also has a more creative application when I use it to make legs for furniture or even decorative moldings.

The jointer is not difficult to use or maintain, but it requires skill to adjust the machine and change the knives. But like all tools, mastering the jointer takes practice and concentration.

- *Mark Duginske*

Mark Duginske is the author of several books on woodworking tools and techniques. He works as a cabinetmaker in Wausau, Wisconsin.

Maintaining Stationary Tools

The precision and consistency we expect from stationary woodworking machines are only possible if the equipment is kept clean and finely tuned. Whether you have a cantankerous old band saw that needs to be cajoled into making a straight cut, or a brand-new radial arm saw that has slipped out of alignment on the way from the factory, learning how to adjust your stationary machines properly will improve the results and increase your pleasure from them.

Many woodworkers are apprehensive about exploring the nuts and bolts of their tools, and many owner's manuals do not encourage tinkering. However, most stationary power tools are quite simple in their design and construction. Taking the top off a table saw, which sounds like a major operation, is fairly simple to do and it can quickly reveal how the machine works, and exactly what you should clean, adjust, or tweak to keep it running smoothly.

The chapter that follows presents the major stationary power tools used by woodworkers and explains the basic maintenance and troubleshooting procedures for each one. Some of these tasks, such as checking belts, cleaning switches, and keeping tabletops clean (*page15*), apply to most of the tools. Other maintenance tasks are specific to the design of a particular machine, such as cleaning and adjusting the blade height and tilt mechanisms on a table saw (*page 62*), or fixing an unbalanced band saw wheel (*page 100*).

Knowing how to tune up your stationary tools will not only give you a deeper understanding of how they work; it will also provide you with a list of things to check when shopping for used models. Is a jointer's fence square? How much runout does a drill press chuck have? Does the miter gauge of a table saw slide smoothly?

Many woodworkers tune up their stationary tools just before the start of a major project. This can be difficult to schedule if you are one of those woodworkers who has many projects on the go. In such cases, it is a good idea to devote a little time periodically to maintaining your stationary tools. That way, every project will benefit from the best your tools can give.

A magnetic-base dial indicator checks the spindle of a drill press for runout—the amount of wobble that the spindle would transmit to a bit or accessory. For accurate drilling, the runout should not exceed 0.005 inch.

With the wheel covers open, a long straightedge confirms that the wheels of a band saw are parallel to each other and in the same vertical plane. As shown in the photo at left, the straightedge should rest flush against the top and bottom of each wheel.

Woodworking Machines

Maintaining Stationary Tools

Back to Basics

Basic Stationary Tool Maintenance

Drive belts transmit power from the motor to the moving parts in many stationary power tools, including the jointer, disc sander, planer, and table saw. In high-torque tools such as the table saw shown in the photo at right, three belts are used to drive the arbor. Any drive belt that is cracked or worn extensively should be replaced.

Checking Drive Belts

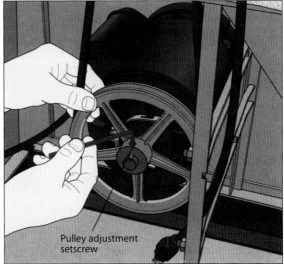

Pulley adjustment setscrew

Checking belt tension
Too much belt tension can strain a stationary tool's motor bearings, while too little tension often leads to slippage and excessive wear. To check drive belt tension on the jointer shown above, unplug the tool and remove the panel covering the belt. Then pinch the belt between the pulleys with one hand *(above, left)*. The amount of deflection will vary with the tool; as a rule of thumb, the belt should flex $1/32$ inch for every inch of span between pulleys. If there is too little or too much tension, adjust it following the manufacturer's instructions. For smooth operation, the pulleys should be aligned; if they are not, loosen the adjustment setscrew on the motor pulley with a hex wrench *(above, right)*, and slide the pulley in line with the other pulley.

Maintaining Tabletops

Cleaning a stationary machine tabletop
To keep stock running smoothly, clean the tabletop frequently, wiping off any pitch or gum deposits with a rag and mineral spirits. Remove any rust or pitting with fine steel wool and penetrating oil *(left)*; then, wipe off any residue and sand the area with fine sandpaper. A coat of paste rubbed on and then buffed will make pushing wood into the cutting edge much less tiring.

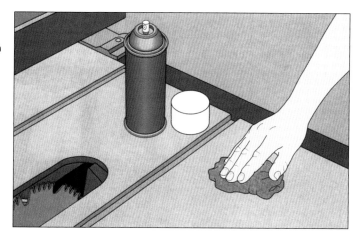

Maintaining Switches

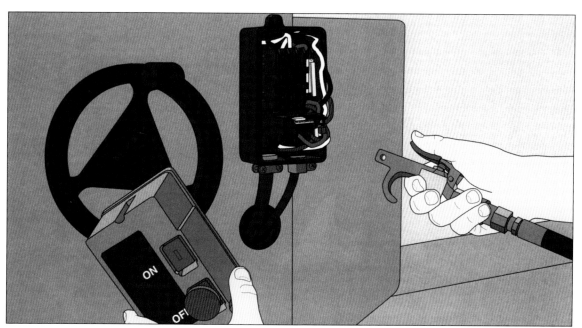

Cleaning a power switch
The switches on stationary tools can become clogged, causing the switch to stick or even preventing it from operating. If the switch sticks, unplug the tool, remove the switch cover and clean the switch immediately. To prevent such problems, periodically clean out the switch by blowing compressed air into it *(above)*.

Portable Power Tools

Whatever their price range or list of features, all portable power tools will work better and last longer if they are cared for properly. At its most basic, preventive maintenance is easy to do and takes no more than a few minutes. At the end of your work day, for example, get in the habit of cleaning dust and dirt from your tools. Refer to the schedules on page 18 for additional maintenance ideas. When you buy a new tool, register the warranty and file the owner's manual in a convenient place and follow all the operating and maintenance instructions suggested by the manufacturer. Owner's manuals typically include troubleshooting guides to help users recognize and handle malfunctions. Keep your tool's original packaging should you need to return an item for servicing.

Because portable tools are electrically powered, caring for them is as much a matter of safety as of performance. Today's power tools are designed to insulate the user from electrical shock, but any tool that develops an electrical problem can be hazardous. This chapter provides illustrations of the portable power tools commonly used in woodworking with cutaway views of their principal electrical and mechanical components.

The drawings are designed to help show you where these parts are typically located and recognize where a tool may have a problem.

Fortunately, the parts of a power tool that endure the most abuse and most often suffer damage are those that are also the easiest to access: the plugs, power cords, motor brushes, and on/off switches. As shown beginning on page 26, these components can be replaced easily and inexpensively. Before undertaking a repair, however, check whether the tool is still covered by the manufacturer's warranty. Opening up a tool that is still under warranty will usually void the guarantee.

The decision to repair other parts of a portable power tool, such as the motor and motor bearing, for example, depends on a number of factors, including your own abilities. The age and value of a tool is also a consideration. The most worthwhile remedy for a 20-year-old drill with a burned-out motor may be a new drill rather than a new motor.

A belt sander in a commercial stand is paired with a plywood truing jig to correct a router's out-of-round sub-base, which can produce imprecise cuts. To correct the problem, install a centering pin in the router, drill a hole in the jig to hold the pin, and turn on the sander. Then slowly rotate the sub-base against the belt until it is perfectly round.

A combination square confirms that the blade of a circular saw is perpendicular to the tool's base plate. All power saws rely on this alignment for accurate cuts.

Woodworking Machines

Portable Power Tools

*Back to **Basics***

Maintenance Tips and Schedules

There are no industry-wide standards for servicing portable power tools designed for the home shop. Manufacturers of industrial-use power tools issue maintenance schedules for their products, but these tools typically undergo heavier use than the average home workshop tool. For industrial tools, servicing is usually scheduled every 100 hours of use and includes a complete overhaul. Brushes are replaced, bearings are cleaned and lubricated (or replaced), and the wiring, motor, and other electrical components are checked and, if necessary, repaired.

For the typical power tool in the home shop, however, maintenance schedules and requirements are less clearcut. Much depends on how a tool is used. A circular saw used by the weekend woodworker to cut the occasional plank will obviously require less attention than one used by a busy carpenter or cabinetmaker.

The chart below lists the checks that should be made on many portable power tools. The tasks listed are straightforward and can be done in a matter of minutes. How often you perform these checks will depend on your own needs and circumstances. As a rule of thumb, any tool that does not perform the way it is designed to should be investigated. You can do the work yourself, but be aware that troubleshooting electrical problems in a power tool requires specialized equipment as well as a sound knowledge of how to use it. If you are uncomfortable working with electricity, take the tool to an authorized service center for repair.

While tools made a few decades ago can be completely disassembled, many recent models feature internal components that are factory-sealed and virtually inaccessible. In some tools, for example, the bearings are mechanically pressed onto the motor spindle. Attempting to separate the bearing from the motor in such tools without the right instrument will destroy the bearing.

To get the most from your tools and keep repairs to a minimum, refer to the tips listed below. Read your owner's manual before using a tool to make certain you can operate it properly. And never try to use a tool for a task for which it is not designed.

Tool	Maintenance
Router	Check the collet for play and run out *(page 21)*
	Clean the collet and spindle
	Ensure that the sub-base is smooth and free of damage
Saber Saw	Check the guide rollers and blade supports for wear
	Check the blade clamp
	Check that base is square to blade
Plate Joiner	Check the plunge mechanism for play
	Check the blade and spindle for wear
	Inspect the pins and glides
	Inspect the drive belt
Electric Drill	Check the chuck bearing for play
	Inspect the chuck for wear
Belt Sander	Check the steel platen and cork pad for wear
	Check the drive belt
	Check the end roller for damage or excessive play
	Inspect the condition of the rubber on the drive roller
Circular Saw	Lubricate the gears
	Check the arbor bearings
	Check the guard return springs
	Check blade alignment
Orbital Sander	Check the pad for wear or splitting
	Check the eccentric bearing (on random-orbit sander)
	Check the pad support

Anatomy of an Electric Drill

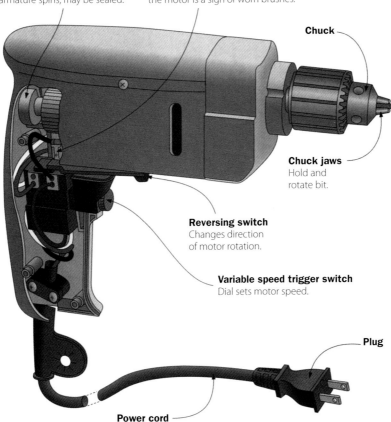

Motor bearing
Located at end of motor shaft to reduce friction as motor armature spins; may be sealed.

Brush assembly
A spring-loaded carbon rod encased in a housing; conducts current to the motor armature. Sparks flying from the motor is a sign of worn brushes.

Chuck

Chuck jaws
Hold and rotate bit.

Reversing switch
Changes direction of motor rotation.

Variable speed trigger switch
Dial sets motor speed.

Plug

Power cord

Anatomy of a Router

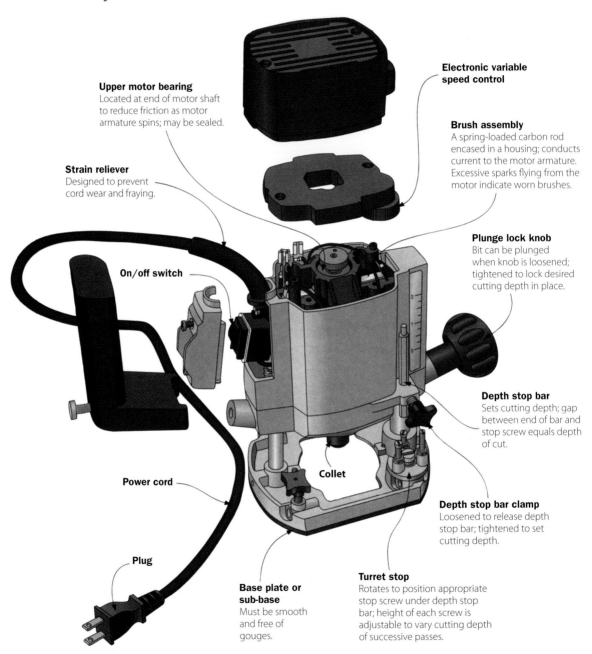

Checking the Collet for Runout

Using a dial indicator and a magnetic base
Install a centering pin in the router as you would a bit and set the tool upside down on a metal surface, such as a table saw. Connect a dial indicator to a magnetic base and place the base next to the router. Turn on the magnet and position the router so the centering pin contacts the plunger of the dial indicator. Calibrate the dial indicator to zero following the manufacturer's instructions. Then turn the shaft of the router by hand to rotate the centering pin *(right)*. The dial indicator will register collet runout—the amount of wobble that the collet is causing the bit. If the runout exceeds 0.005 inch, replace the collet.

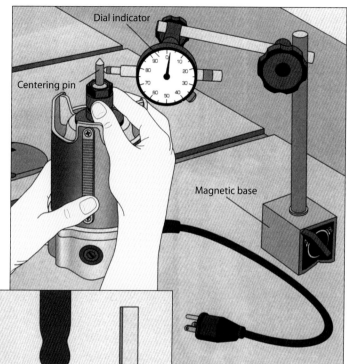

Using a feeler gauge
If you do not have a dial indicator, you can test for collet runout with a feeler gauge and a straight hardwood block. With the centering pin in the collet and the router upside down on a work surface, clamp the block lightly to the tool's sub-base so the piece of wood touches the pin. Turn the router shaft by hand; any runout will cause the centering pin to move the block. Then use a feeler gauge to measure any gap between the pin and the block *(left)*. If the gap exceeds 0.005 inch, replace the collet.

Anatomy of a Saber Saw

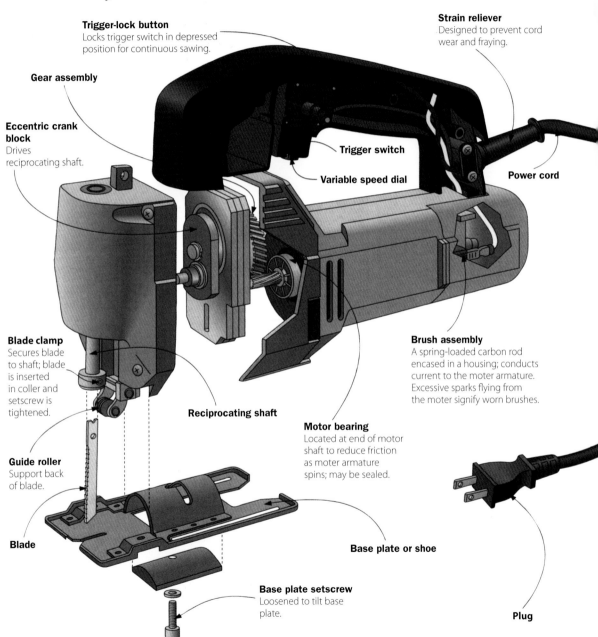

Woodworking Machines

Squaring the Blade

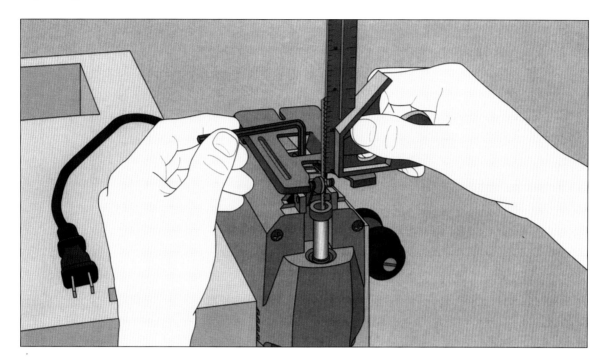

Checking the blade angle
Square a saber saw blade each time you install a new blade. Unplug the saw, then secure it upside down in a bench vise as shown above. Use a combination square to check whether the blade is square with the base plate. If not, loosen the base plate setscrew with a hex wrench and tilt the plate until the blade butts flush against the square. Then tighten the setscrew.

Shop Tip

Extending blade life
If most of the stock you cut is ¾ inch or thinner, the top third of your blade will be the only portion showing wear. To make better use of the full length of the cutting edge, install an auxiliary shoe on the base plate of the saw once the top third of a blade begins to dull. To make the shoe, cut a piece of ½-inch plywood the same length as the base plate and slightly wider. Hold the wood against the plate and mark the outline of the notch cut out for the blade. Saw out the notch and cut a slot for the blade. Screw the auxiliary shoe in place, making sure that the back of the blade fits in the slot. (If the blade is not supported, it may wander and break when you are cutting.)

Anatomy of a Plate Joiner

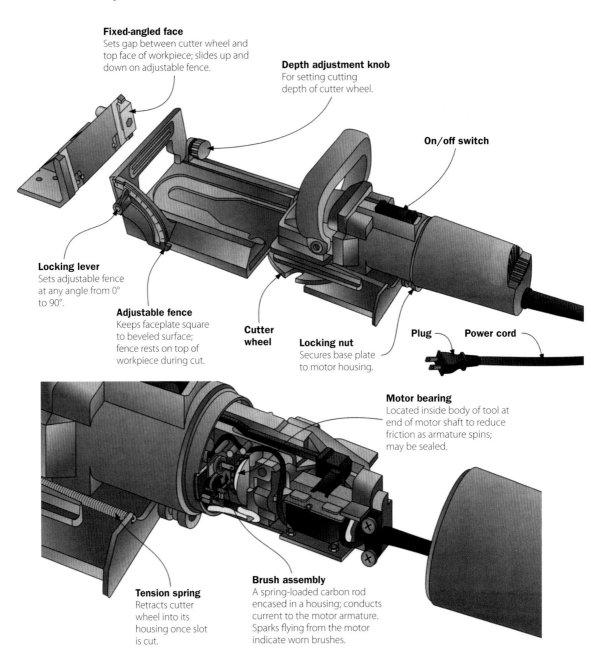

Anatomy of a Circular Saw

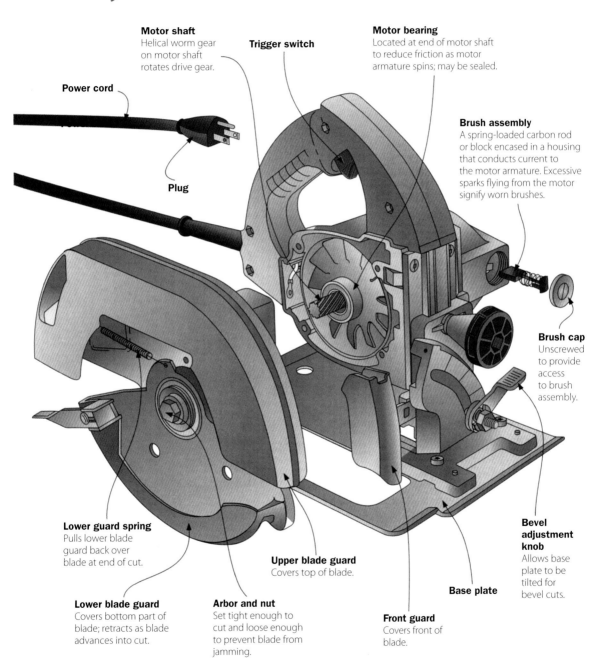

Repairing Portable Power Tools

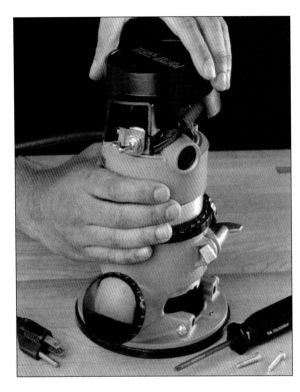

The cap housing of a router is lifted off the body of the tool, revealing the wiring connections for the toggle switch inside. As shown in the photo at left, gaining access to the internal components of most portable power tools is simply a matter of loosening the retaining screws that secure the tool housing to the body. The cap providing access to the brush assembly for this router is located on the side of the tool body to the right of the switch.

Replacing a Brush Assembly

Removing and installing a brush assembly
Brushes are spring-loaded carbon rods or blocks that conduct electricity to the rotating armature of a power tool's motor. Over time, brushes wear or become damaged. You can access the assembly by unscrewing the brush cap on the tool body—normally a plastic cap roughly the size of a dime. If there is no brush cap, you will have to remove the motor housing to access the brushes. Once you have located the brush assembly, carefully lift it out of the tool. To test the brush, push on it to check the spring. If the spring is damaged or the brush is pitted, uneven, or worn shorter than its width, you will need to replace the assembly. Some brushes are marked with a wear line. Buy a replacement at an authorized service center for your brand of tool. To reinstall a brush assembly, fit it into position in the tool *(right)*. Then insert and tighten the brush cap or reattach the motor housing.

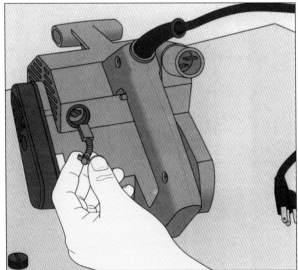

Replacing a Switch

Removing the old switch

Set the tool on a work surface, making sure that it is unplugged. For the router shown at left, remove the cap housing to expose the switch mechanism. Loosen the switch retaining nut and screws, then disconnect the wires securing the mechanism to the tool. If the wires are connected by wire caps, simply loosen the caps *(right)* and untwist the wires. If the connections are soldered, snip the connections with pliers. Use short strips of masking tape to label the wires to help you reconnect them properly.

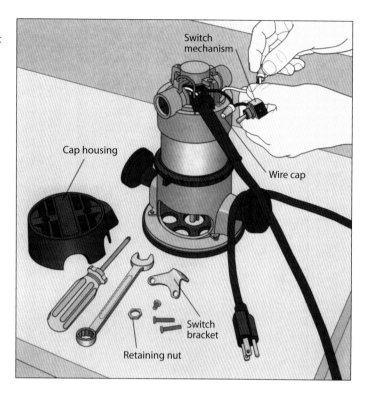

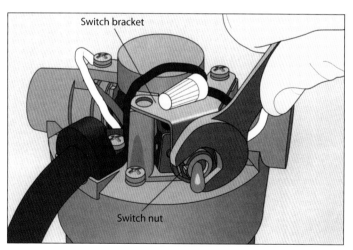

Installing the replacement switch

Buy a replacement switch at an authorized service center, noting the model and serial numbers of your tool. Connect the new switch to the wires in the tool housing, reversing the steps you took to take out the old one. Remove the masking tape strips from the wires, twist the wire ends from the tool and switch together, and screw a wire cap onto each connection to secure and insulate it. Fit the switch into position in the tool housing, screw the switch bracket in place, and tighten the switch retaining nut with a wrench *(left)*. Replace the tool's cap housing.

Replacing a Power Cord

Accessing the cord's wire terminals
The wire terminals connecting a tool's power cord to the switch mechanism are contained within the motor housing. For the sander shown at right, reach the terminals by removing the auxiliary handle and loosening the screws securing the main handle to the tool body. Remove the handle to expose the wire terminals.

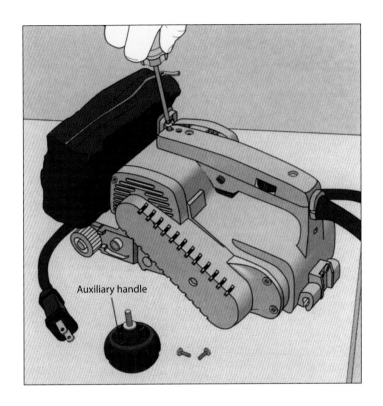

Disconnecting the old power cord
On a power cord with a two-prong plug, there are usually two wires from the cord connected to wire terminal screws in the tool housing. Unscrew the plug retaining bracket securing the cord to the tool housing, loosen the terminal screws *(left)*, and carefully remove the power cord's wire ends from the terminals. Use strips of masking tape to label each terminal to help you attach the wire ends of the new cord to the appropriate terminals.

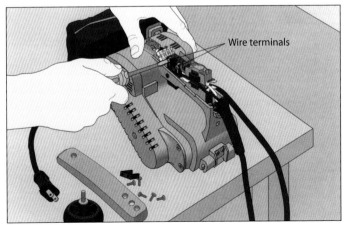

Preparing the replacement power cord

Buy a replacement power cord at a hardware store or an authorized service center, making sure it has the same specifications as the original cord. The wire ends of new power cords are usually covered to the end with jacketing and insulation. To prepare the cord for installation, use a knife to cut away a few inches of the jacket covering the two wires. Then strip off about ½ inch of the plastic insulation around the wires, exposing the ends. You can also use wire strippers for this task. Avoid cutting into the metal wire; if you sever any of the strands, snip off the damaged section and remove more insulation to uncover a fresh section. Use needle-nose pliers to carefully twist the wire strands snugly together *(right)*, then bend the wire ends into semicircles that will hug the terminals in the tool housing. Place the wire around the screw clockwise from the left side, so it will wrap around as the screw is tightened.

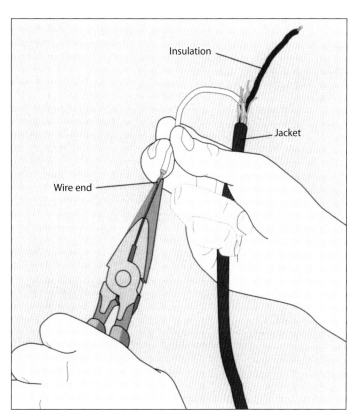

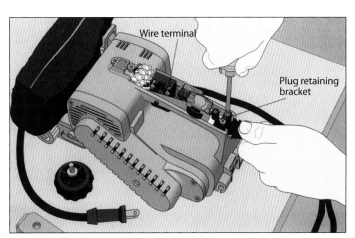

Installing the replacement cord

Hook the wire ends around the terminals in the tool housing, making sure to attach each wire to the appropriate screw. Remove the masking tape. Holding the power cord in position, screw the cord retaining bracket in place *(left)*, then reinstall the handles on the tool body.

Sharpening Blades and Bits

Like any cutting or shaping tool, a power tool with a dull blade or bit cannot perform well. A dull drill bit will tend to skate off a workpiece, rather than biting cleanly into the wood. A saw blade or router bit with blunted cutting edges may burn stock. And wood that is surfaced by a jointer or planer with unsharpened knives may be difficult to glue up or finish.

In addition to cutting and shaping properly, well-sharpened blades and bits offer other benefits, including reduced wear and tear on motors, less operator fatigue, and longer life for the blades and bits themselves. Manufacturers of power tool blades and bits generally recommend sending their products to a professional sharpening service. However, the job can often be done in the workshop. This chapter will show you how to sharpen a wide variety of power tool blades and bits, from router bits and shaper cutters *(page 34)* to jointer and planer knives *(page 49)*. In a pinch, even a broken band saw blade can be soldered together *(page 46)*.

Still, there are times when you should turn to a professional, particularly if blades and bits have chipped edges or have lost their temper as a result of overgrinding. Some router bits also must be precisely balanced, something that is difficult to achieve in the shop. As a rule of thumb, it is a good idea to send out your bits and blades to a sharpening service periodically, or every second time they need a major sharpening. Once you have sharpened an edge properly, it should last for a long time—the occasional honing is all that it takes to maintain it.

The pages that follow cover the basic techniques for sharpening power tool blades and bits in the shop. With a little practice and the right accessories, you can keep the cutting edges of your blades and bits razor-sharp. But remember that a keen edge always starts with the quality of the steel itself; for long life and ease of sharpening, always choose bits and blades made from the best steel.

Designed to replace the metal guide blocks supplied with most band saws, heat-resistant guide blocks can help prolong blade life. Made from a graphite-impregnated resin that is its own lubricant, these nonmetallic blocks last longer than metal blocks and can be set closer to the blade, allowing more accurate and controlled cuts.

A twist bit is sharpened on a bench grinder with the help of a commercial grinding jig that holds the bit at the proper angle. Originally designed for the metal-working industry, twist bits took their place in woodworking as the use of power tools grew. They need periodic sharpening to drill holes cleanly and accurately.

Woodworking **Machines**

Sharpening Blades and Bits

Back to **Basics**

A Gallery of Blades and Bits

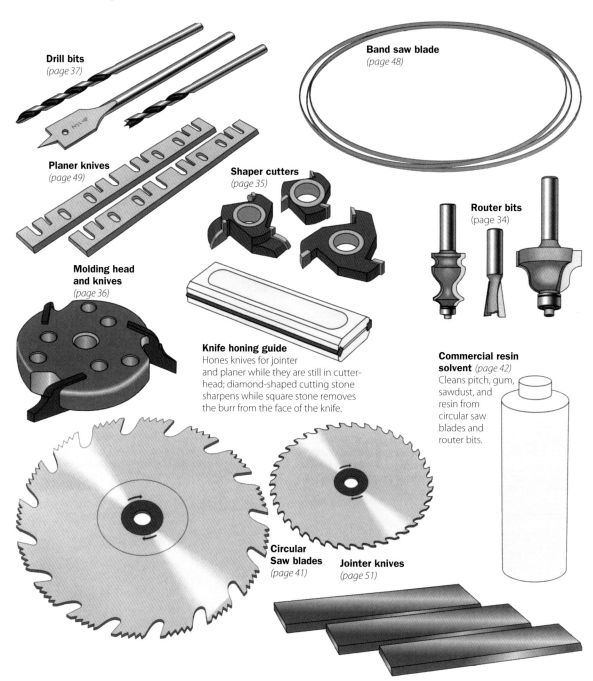

Drill bits
(page 37)

Band saw blade
(page 48)

Planer knives
(page 49)

Shaper cutters
(page 35)

Router bits
(page 34)

Molding head and knives
(page 36)

Knife honing guide
Hones knives for jointer and planer while they are still in cutterhead; diamond-shaped cutting stone sharpens while square stone removes the burr from the face of the knife.

Commercial resin solvent (page 42)
Cleans pitch, gum, sawdust, and resin from circular saw blades and router bits.

Circular Saw blades
(page 41)

Jointer knives
(page 51)

Woodworking Machines

Tools and Accessories for Sharpening

Drill bit grinding attachment *(page 37)*
Holds ⅛- to ¾-inch-diameter twist bits for sharpening; mounted to work surface and used with a bench grinder.

Router bit sharpener
A boron-carbide stone used to sharpen carbon steel, high-speed steel, and carbide-tipped router bits; gives a finer finish than diamond files of equal grit. Handle features magnifying lens for checking sharpness.

Drill bit-sharpening jig *(page 37)*
Powered by an electric drill, this jig sharpens high-speed steel twist bits and carbide masonry bits up to ½ inch in diameter; holder secures bit at proper depth and angle against sharpening stone inside jig.

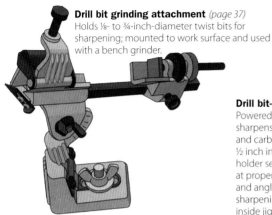

Circular saw blade-setting jig *(page 43)*
Clamped in bench vise to joint and set the teeth of circular saws up to 12 inches in diameter. Blade is locked in jig and rotated against file to joint teeth; teeth are set by tapping them with a hammer against mandrel.

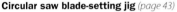

Knife-setting jigs *(page 55)*
Magnetic jig used to hold jointer or planer knives at the correct height for installation in the machine. Jigs for planer (below) are used in pairs for knives up to 20 inches long; jig for jointer *(right)* sets knives up to 8 inches in length, and can be extended with a third bar for knives up to 14 inches long.

Jointer/planer-knife sharpening jig *(page 52)*
Used to sharpen jointer and planer knives; knife is clamped in jig and rear screw adjusts to hold knife at proper angle against a bench stone.

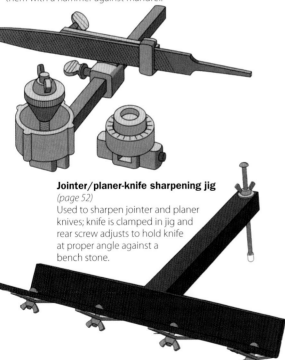

Circular saw blade-sharpening jig *(page 43)*
Mounted on workbench to sharpen circular saw blades after grinding and setting; blade is held in jig while taper file is drawn across the teeth at the proper pitch and angle.

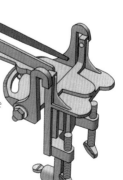

Sharpening Blades and Bits

Back to Basics

33

Router Bits and Shaper Cutters

Secured in a bench vise, one of the cutting edges of a shaper cutter receives its final sharpening with a fine diamond hone. The process is a two-step operation, beginning with a medium hone *(left)*. Because they operate at high speeds, dull router bits and shaper cutters overheat quickly. Cutters that are properly sharpened make smoother, more accurate cuts.

Sharpening a Non-Piloted Router Bit

Sharpening the inside faces
Clean any pitch, gum, or sawdust off the bit with a commercial resin solvent *(page 42)*, then use a ceramic or diamond sharpening file to hone the inside faces of the bit's cutting edges. A coarse-grit file is best if a lot of material needs to be removed; use a finer-grit file for a light touch-up. Holding the inside face of one cutting edge flat against the abrasive surface, rub it back and forth *(right)*. Repeat with the other cutting edge. Hone both inside faces equally to maintain the balance of the bit. Take care not to file the bevel behind the cutting edge.

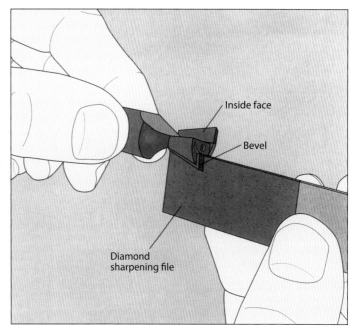

Sharpening a Piloted Router Bit

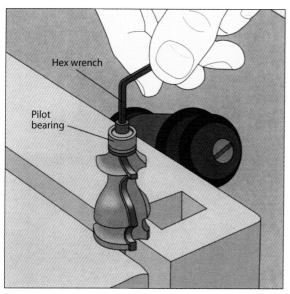

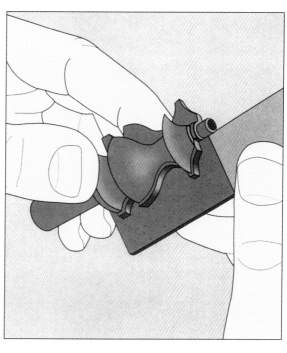

Removing the pilot bearing
Before you can sharpen a piloted router bit, you need to remove the pilot bearing. Use a hex wrench to loosen the bearing *(above)*.

Sharpening the bit
Sharpen the bit with a ceramic or diamond sharpening file as you would a non-piloted bit *(page 34);* then re-install the bearing with the hex wrench. If the bearing does not rotate smoothly, spray a little bearing lubricant on it. If it is worn out or damaged, replace it.

Shop Tip

A storage rack for shaper cutters
Shaper cutters are often sold in cumbersome packaging that can contribute to clutter. Organize your shaper bits with a shop-made storage rack like the one shown here. The rack will keep the cutters visible and accessible. Drill a series of holes in a board and glue dowels in the holes to hold the cutters. To prevent the cutting edges from nicking each other, use your largest-diameter cutter as a guide to spacing the dowel holes. If you plan to hang the rack on a wall, bore the holes at a slight angle so that the cutters will not slip off the dowels.

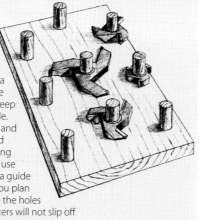

Molding Knives

Sharpening Molding Knives

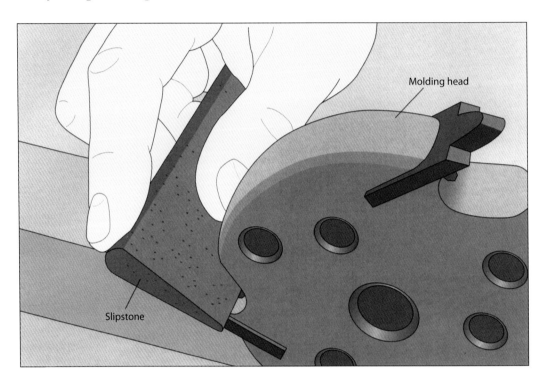

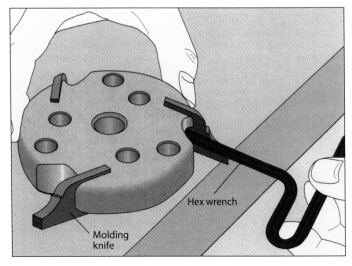

Sharpening molding knives
The cutting edges of table saw or radial arm saw molding knives are easy to touch up or sharpen while they are mounted in the molding head. Clamp the head in a bench vise with one of the knives clear of the bench, then use a slipstone *(above)* to hone its inside face as you would a router bit *(page 34)*. Reposition the head in the vise to hone the remaining knives. Use the same number of strokes to hone each knife so that you remove an equal amount of metal from them all, and maintain their identical shapes and weights. An alternative method involves removing the knives with a hex wrench *(left)* and sharpening them on a flat oilstone.

Drill Bits

Sharpening Twist Bits

Using a bench grinder
Holding the bit between the index finger and thumb of one hand, set it on the grinder's tool rest and advance it toward the wheel until your index finger contacts the tool rest. Tilt the shaft of the bit down and to the left so that one of the cutting edges, or lips, is square to the wheel *(right)*. Rotate the bit clockwise to grind the lip evenly. Periodically check the angle of the cutting edge, as shown in the photo at right, and try to maintain the angle at about 60°. Repeat for the second cutting edge. To keep bits sharp, use them at the speed recommended by the manufacturer. Wipe them occasionally with oil to prevent rust.

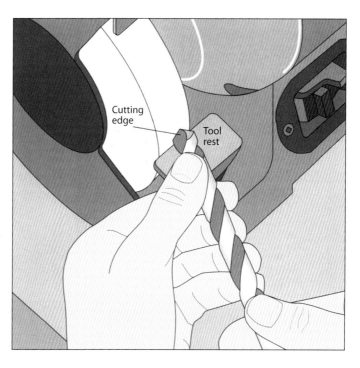

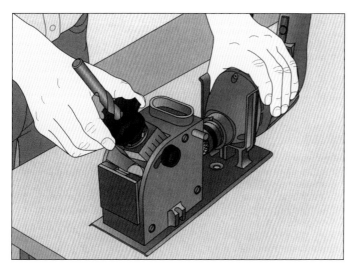

Using a commercial jig
Set up the jig following the manufacturer's instructions. For the model shown, secure an electric drill to the jig; the drill will rotate the sharpening stone inside the device. Adjust the angle block to the appropriate angle for the bit to be sharpened and insert the bit in the depth gauge. The gauge will enable you to secure the bit at the correct height in the holder. Fit the bit holder over the bit and then use it to remove the bit from the gauge. Now secure the bit and holder to the angle block. Turn on the drill and, holding it steady, slowly rotate the bit holder a full 360° against the stone inside the jig *(left)*. Apply light pressure; too much force will overheat the bit.

Sharpening Forstner Bits

Grinding the inside bevel
To touch up a Forstner bit, true the top edge of the bit's rim with a file, removing any nicks. If the beveled edges of the cutting spurs inside the rim are uneven, grind them using an electric drill fitted with a rotary grinding attachment. Secure the bit in a bench vise as shown and grind the edges until they are all uniform *(right)*.

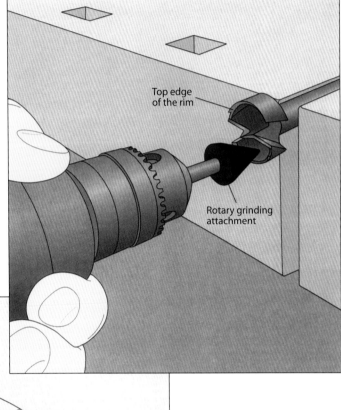

Sharpening the chip lifters
Use a single-cut mill bastard file to lightly file the inside faces of the cutters. Hold the file flat against one of the cutters—also known as chip lifters—and make a few strokes along the surface *(left)*. Repeat with the other cutter. Finish the job by honing the beveled edges inside the rim with a slipstone.

Honing Multi-Spur Bits

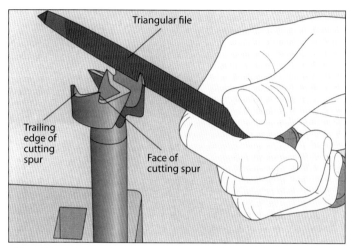

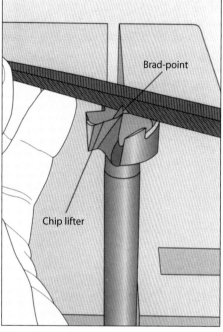

Filing the cutting spurs
Secure the bit upright in a bench vise and use a triangular file to hone the leading edge, or face, of each spur *(above)*. File with each push stroke, towards the bit's brad point, tilting the handle of the file down slightly. Then file the trailing edge, or back, of each spur the same way. File all the spurs by the same amount so that they remain at the same height. Make sure you do not over-file the cutting spurs; they are designed to be 1/32 inch longer than the chip lifters.

Filing the brad point
File the chip lifters as you would those of a Forstner bit *(page 38)*. Then, file the brad-point until it is sharp *(above)*.

Sharpening Brad-Point Bits

Filing the chip lifters
Clamp the bit upright in a bench vise and file the inside faces of the two chip lifters as you would those of a Forstner bit *(page 38)*. For a brad-point bit, however, use a triangular needle file *(right)*, honing until each cutting edge is sharp and each chip lifter is flat.

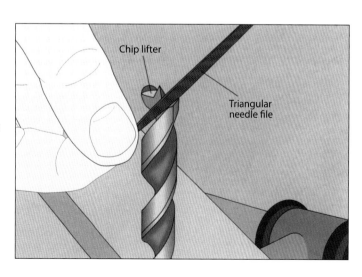

Filing the cutting spurs
Use the needle file to hone the inside faces of the bit's two cutting spurs. Hold the tool with both hands and file towards the brad-point until each spur is sharp *(right)*.

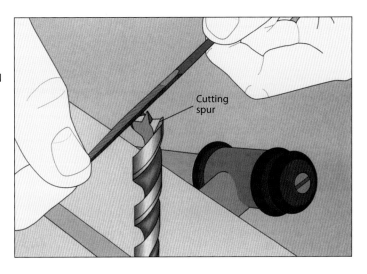

Honing Spade Bits

Filing a spade bit
Secure the bit in a bench vise and use a smooth single-cut mill bastard file to touch up the two cutting edges. File on the push stroke *(left)*, tilting the handle down slightly to match the angle of the cutting edges; between 5° and 10° is typical. Then touch up the cutting edges on either side of the point the same way *(inset)*, taking care not to alter its taper. Do not remove too much metal at the base of the point, as this will weaken the bit.

Circular Saw Blades

Changing Table Saw Blades

Removing a blade
Working at the front of the table, remove the insert and wedge a piece of scrap wood under a blade tooth to prevent the blade from turning. Use the wrench supplied with the saw to loosen the arbor nut *(right)*. (Most table saw arbors have reverse threads; the nut is loosened in a clockwise direction.) Finish loosening the nut by hand, making sure that it does not fall into the machine. Carefully lift the blade and washer off the arbor. Carbide-tipped blades are best sharpened professionally; but high-speed steel models can be sharpened in the shop *(page 43)*. A worn or damaged blade should be discarded and replaced.

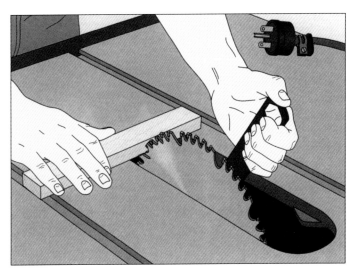

Installing a blade
Slide the blade onto the arbor with its teeth pointing in the direction of blade rotation (toward the front of the table). Insert the flange and nut and start tightening by hand. To finish tightening, grip the saw blade with a rag and use the wrench supplied with the saw *(left)*. Do not use a piece of wood as a wedge, as this could result in overtightening the nut. Always unplug tools before beginning any work.

Sharpening a Band Saw Blade

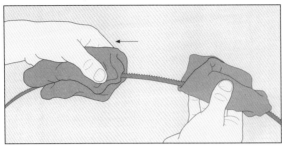

Cleaning the blade
Before sharpening a band saw blade, remove sawdust and wood chips from it. Make sure you release the blade tension first before slipping the blade off the wheels. Then, holding the blade between two clean rags *(above)*, pull it away in the direction opposite its normal rotation to avoid snagging the cutting edges in the material.

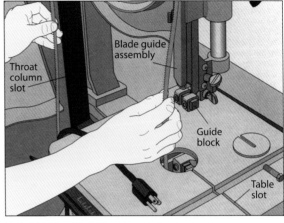

Installing the blade for sharpening
You can sharpen a band saw blade either on a bench vise or on the machine. To install the blade on the band saw for sharpening, mount it with the teeth pointing in the direction opposite their cutting position—that is, facing up instead of down. Turn the blade inside out and guide it through the table slot *(above)*, holding it with the teeth facing you and pointing up. Slip the blade between the guide blocks and in the throat column slot, then center it on the wheels. Make sure the blade guide assembly is raised as high above the table as it will go.

Setting the blade
If the teeth need to be set, adjust a commercial saw set to the same number of teeth per inch as the band saw blade. Secure the blade in a handscrew and clamp the handscrew to the saw table. Starting at the handscrew-end of the blade, position the first tooth that is bent to the right between the anvil and punch block of the saw set and squeeze the handle to set the tooth *(right)*. Work your way up to the guide assembly, setting all the teeth that are bent to the right. Then turn the saw set over and repeat for the leftward-bent teeth. Continue setting all the blade teeth section by section. To ensure you do not omit any teeth, mark each section you work on with chalk.

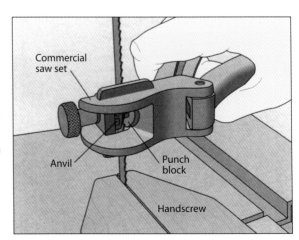

Sharpening the blade

Sharpen the teeth the same way you set them, working on one blade section at a time. Hold a triangular file at a 90° angle to the blade and sharpen each tooth that is set to the right, guiding the file in the same direction that the tooth is set *(right)*. Then sharpen the leftward-bent teeth the same way. Use the same number of strokes on each tooth. Once all the teeth have been sharpened, remove the blade, turn it inside out and reinstall it for cutting, with the teeth pointing down. Tension and track the blade.

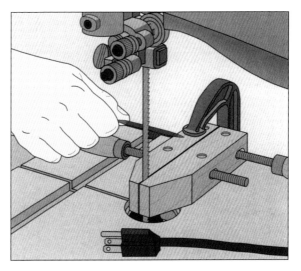

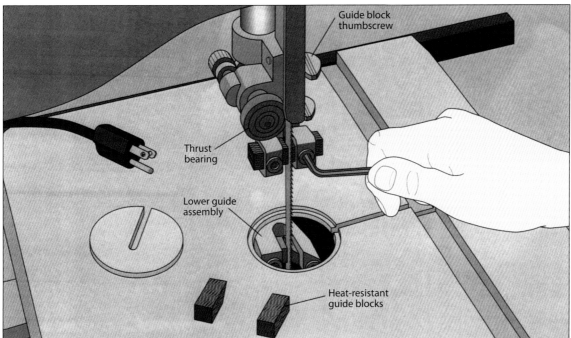

Installing heat-resistant guide blocks

Replacing your band saw's standard guide blocks with heat-resistant blocks will lengthen blade life and promote more accurate and controlled cuts. Remove the original blocks by using a hex wrench to loosen the setscrews securing them to the upper guide assembly *(above)*. Slip out the old blocks and insert the replacements. Pinch the blocks together with your thumb and index finger until they almost touch the blade.

(You can also use a slip of paper to set the space between the guide blocks and the blade). Tighten the setscrews. The front edges of the guide blocks should be just behind the blade gullets. To reposition the blocks, loosen their thumbscrew and turn their adjustment knob to advance or retract the blocks. Tighten the thumbscrew and repeat the process for the guide assembly located below the table.

Repairing a Broken Band Saw Blade

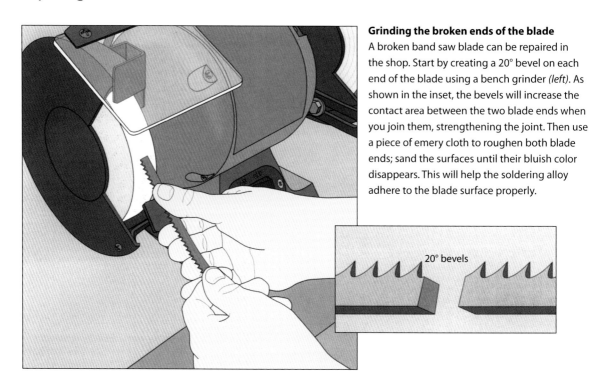

Grinding the broken ends of the blade
A broken band saw blade can be repaired in the shop. Start by creating a 20° bevel on each end of the blade using a bench grinder *(left)*. As shown in the inset, the bevels will increase the contact area between the two blade ends when you join them, strengthening the joint. Then use a piece of emery cloth to roughen both blade ends; sand the surfaces until their bluish color disappears. This will help the soldering alloy adhere to the blade surface properly.

Setting up the blade in the soldering jig
Secure a commercial soldering jig in a machinist's vise. Next, use a brush to spread flux on the beveled ends of the blade and ½ inch in from each end. Position the blade in the jig so the two beveled ends are in contact *(right)*. Make sure the blade is tight and straight in the jig.

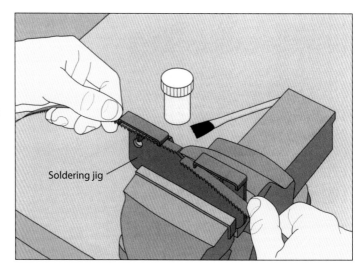

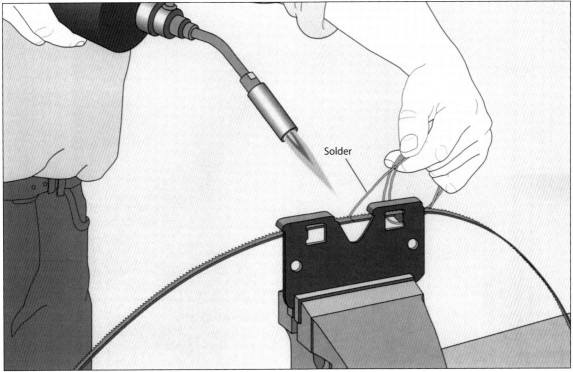

Soldering the blade ends
Heat the joint with a propane torch, then unroll a length of the solder and touch the tip to the joint—not to the flame. Continue heating the joint *(above)* until the solder covers the joint completely. Turn off the torch and let the joint cool.

Filing the joint
Once the joint has cooled, remove the blade from the jig and wash off the flux with warm water. If there is an excess of solder on the blade, file it off carefully with a single-cut bastard mill file *(left)* until the joint is no thicker than the rest of the blade. If the joint separates, reheat it to melt the solder, pull it apart, and repeat.

Folding and Storing a Band Saw Blade

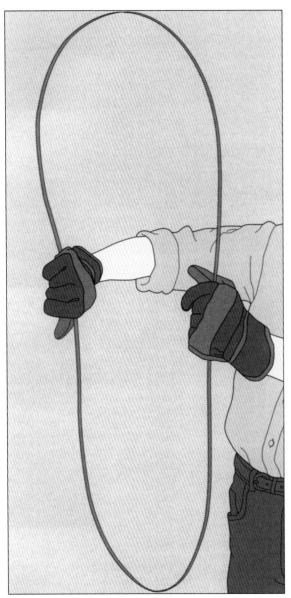

Holding the blade
Before storing a band saw blade, remove any rust from it with steel wool and wipe it with an oily rag. Then, wearing safety goggles and gloves, grasp the blade with the teeth facing away from you; point your left thumb up and your right thumb down *(above)*.

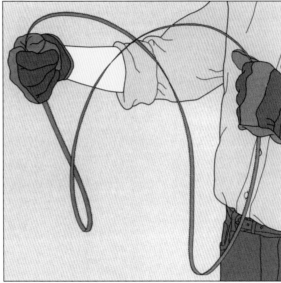

Twisting the blade
Pressing your right thumb firmly against the blade, twist it by pivoting your right hand upward. The blade will begin to form two loops *(above)*.

Coiling the blade
Without pausing or releasing the blade, keep rotating it in the same direction while pivoting your left hand in the opposite direction. The blade will coil again, forming a third loop *(above)*. Secure the blade with string, pipe cleaners, or plastic twist ties.

Jointer and Planer Knives

A pair of magnetic jigs holds a planer knife at the correct height in the cutter-head, allowing the knife to be fixed in place accurately. Such jigs take the guesswork out of the trickiest phase of sharpening planer knives—installing them properly. Periodic sharpening of planer knives is essential. Stock that is surfaced by dull knives is difficult to glue and does not accept finishes well. A similar jig is available for setting jointer knives.

Honing Jointer Knives

Cleaning the knives
Jointer knives can be honed while they are in the cutter-head. Start by cleaning them. Shift the fence away from the tables and move the guard out of the way. Making sure the jointer is unplugged, rotate the cutterhead with a stick until one of the knives is at the highest point in its rotation. Then, holding the cutterhead steady with one hand protected by a rag, use a small brass-bristled brush soaked in solvent to clean the knife *(above)*. Repeat for the other knives.

Aligning the infeed table with the knives

Cut a piece of ¼-inch plywood to the width of the jointer's infeed table and secure it to the table with double-faced tape. The plywood will protect the table from scratches when you hone the knives. Next, unplug the planer and adjust the infeed table so that the beveled edge of the knives is at the same level as the top of the plywood. Set a straight board on the plywood and across the cutterhead and, holding the cutterhead steady with the beveled edge of one knife parallel to the table, lower the infeed table until the bottom of the board contacts the bevel *(left)*. Use a wood shim to wedge the cutterhead in place.

Honing the knives

Slide a combination stone evenly across the beveled edge of the knife *(right)*. Move the stone with a side-to-side motion until the bevel is flat and sharp, avoiding contact with the cutterhead. Repeat the process to hone the remaining knives.

Sharpening Jointer Knives

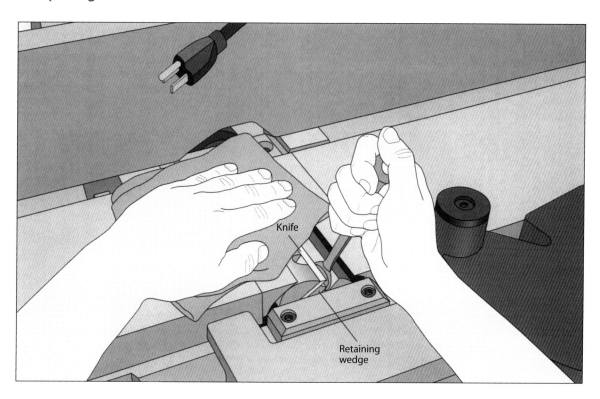

Shop Tip

Shifting knives for longer life
To prolong the life of a set of jointer knives that have been nicked, loosen the lock screws securing one knife and slide the knife about 1/16 inch in either direction. Tighten the lock screws and carefully rotate the cutter-head by hand to ensure that the knife turns freely. Shifting a knife to one side moves its damaged segment out of alignment with the damage on the other knives, enabling the set to continue cutting smoothly.

Removing the knives
To give jointer knives a full-fledged sharpening, remove them from the cutterhead. Unplug the machine, shift the fence away from the tables, and move the guard out of the way. Use a small wood scrap to rotate the cutterhead until the lock screws securing one of the knives are accessible between the tables. Cover the edge of the knife with a rag to protect your hands, then use a wrench to loosen each screw *(above)*. Lift the knife and the retaining wedge out of the cutterhead.

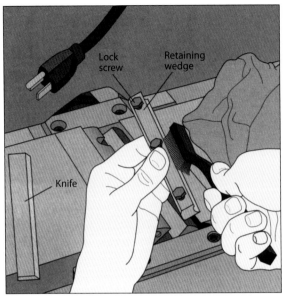
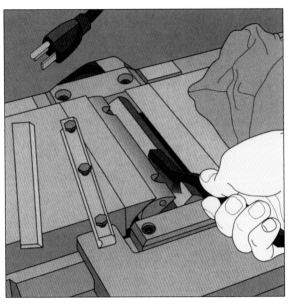

Cleaning the retaining wedge
Unplug the equipment and clean any pitch or gum from the retaining wedge using a brass-bristled brush dipped in solvent *(above, left)*. If the face of the retaining wedge that butts against the knife is pitted or rough, you may have trouble setting the knife height when reinstalling the knife. Flatten the face of the wedge as you would the sole of a plane until it is smooth. Also use the brush to clean the slot in the cutterhead that houses the retaining wedge and knife *(above, right)*.

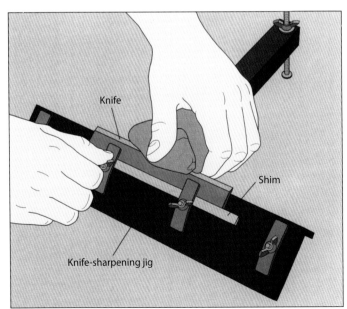

Installing the knife in a sharpening jig
Use a commercial knife-sharpening jig to sharpen the jointer knife. Center the knife in the jig bevel up and clamp it in place by tightening the wing nuts; use a rag to protect your hand *(left)*. Make sure that the blade is parallel with the lip of the jig. If the knife does not extend out far enough from the jig, insert a wood shim between the knife and the jig clamps.

Woodworking Machines

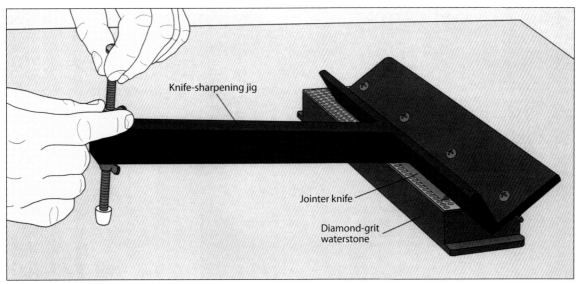

Knife-sharpening jig
Jointer knife
Diamond-grit waterstone

Sharpening the knife
Set a sharpening stone on a flat, smooth work surface; in the illustrations on this page, a diamond-grit waterstone is shown. To adjust the jig so the beveled edge of the jointer knife is flat on the stone, turn the jig over, rest the bevel on the stone, and turn the wing nuts at the other end of the jig *(above)*. Lubricate the stone—in this case with water—and slide the knife back and forth. Holding the knob-end of the jig flat on the work surface and pressing the knife on the stone, move the jig in a figure-eight pattern *(below)*. Continue until the bevel is flat and sharp. Carefully remove the knife from the jig and hone the flat side of the knife to remove any burr formed in the sharpening process.

Reinstalling the knife in the jointer

Insert the retaining wedge in the cutterhead, centering it in the slot with its grooved edge facing up. With the beveled edge of the knife facing the outfeed table, slip it between the retaining wedge and the front edge of the slot, leaving the bevel protruding from the cutterhead.

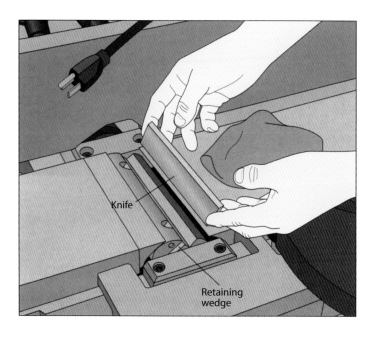

Setting the knife height

Adjust the height of the knife using a commercial jig *(page 52)*, or do the job by hand, as shown at left. Cover the edge of the knife with a rag and partially tighten each lock screw on the retaining wedge. Use a small wooden wedge to rotate the cutterhead until the edge of the knife is at its highest point—also known as Top Dead Center or TDC. Then, holding the cutterhead stationary with a wedge, place a straight hardwood board on the outfeed table so that it extends over the cutterhead. The knife should just brush against the board along the knife's entire length. If not, use a hex wrench to adjust the knife jack screws. Once the knife is at the correct height, tighten the lock screws on the retaining wedge fully, beginning with the one in the center and working out toward the edges. Sharpen and install the remaining knives the same way.

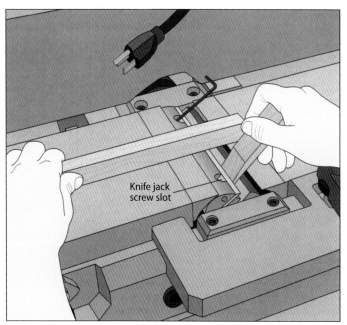

Installing Jointer Knives with a Jig

Using a knife-setting jig
The jig shown at right features magnetic arms that will hold a jointer knife at the correct height while you tighten the retaining wedge lock screws. Insert the knife in the cutterhead and position it at its highest point as you would to install the knife by hand *(page 54)*. Then mark a line on the fence directly above the cutting edge. Position the knife-setting jig on the outfeed table, aligning the reference line on the jig arm with the marked line on the fence, as shown. Mark another line on the fence directly above the second reference line on the jig arm. Remove the jig and extend this line across the outfeed table. (The line will help you quickly position the jig the next time you install a knife.) Reposition the jig on the table, aligning its reference lines with the marked lines on the fence. Then use a wrench to tighten the lock screws *(right)*.

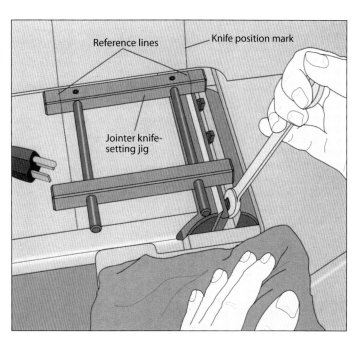

Sharpening Planer Knives

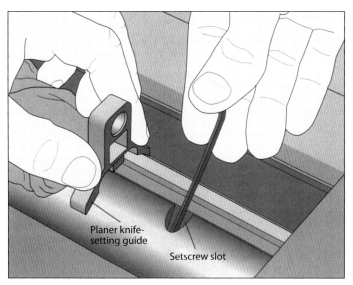

Removing and installing a planer knife
Remove a planer knife from the machine and sharpen it as you would a jointer knife *(page 51)*. To reinstall the knife use the knife-setting guide supplied with the machine or a commercially available model like the one shown on page 49. Place the knife in the planer cutterhead and partially tighten the setscrews. Hold the knife-setting guide beside one of the set-screws so that its two feet are resting on the cutterhead on each side of the opening. Then adjust the setscrew with a hex wrench until the edge of the knife contacts the bottom of the guide *(left)*. Repeat for the remaining setscrews.

Table Saw

Whether you are working with milled boards from a lumberyard, old barn siding, or sheets of 4-by-8 plywood, the table saw is an excellent all-around tool for cutting wood to width (ripping) and length (crosscutting). If the table saw were used for nothing but these two cuts, it would still be a valuable tool. But the saw also accepts a wide variety of blades and accessories, from roller stands that assist with unwieldy panels *(page 74)* to molding heads capable of producing elaborate decorative trim *(page 88)*. And with help from the simple, inexpensive shop-made jigs featured in this chapter, the table saw is also unsurpassed for repeat cuts and also valuable for making such fundamental woodworking joints as the lap, box and open mortise-and-tenon joints *(page 91)*.

The precision and power of a table saw permit a woodworker to make many different cuts with small risk of error. Sawing square and straight with hand tools requires considerable skill and time; but a woodworker who follows the procedures for the table saw outlined in this chapter can produce clean, accurate cuts—consistently, and with relatively little effort.

Table saws are designated according to the blade diameter used. Models are commonly available in 8-, 9-, 10- and 12-inch sizes. The 8- and 10-inch models, however, are clearly the most popular home workshop saws. When choosing a table saw, first consider the type of woodworking you will be doing with it. The fully enclosed stationary saw, like the one pictured on pages 14–15, typically uses a 1.5- to 3-horsepower motor to drive a 10-inch blade. Properly tuned and maintained, it can mill 3-inch stock repeatedly without overheating.

If most of your work is with ¾- or 1-inch-thick stock typically used for cabinetmaking, the open-base contractor's saw is a less expensive alternative. Its 1.5-horsepower motor turns an 8- or 10-inch blade, and the unit can be mounted on a mobile base, providing extra flexibility. In any case, the basic requirement for a table saw—whether for cabinetmaking or general workshop use—is that it must be capable of cutting a 2-by-4 at both 90° and 45°.

For occasional use on light stock or where space is at a premium, the 8¼-inch bench top saw can easily be hauled around the workshop or the job site by one person. When choosing a saw, beware of exaggerated horsepower ratings. Check the motor plate: An honest 1.5-horsepower motor should draw roughly 14 amps at 115 volts; a 3-horsepower motor should draw 14 or 15 amps at 230 volts.

Screwed to the miter gauge, this wood extension facilitates the cutting of box joints for drawers and casework. Such shop-made jigs extend the versatility of the basic table saw.

With a workpiece clamped firmly to a tenoning jig, a woodworker cuts the tenon part of an open mortise-and-tenon joint. The jig is guided by a rail that slides in the saw table's miter gauge slot.

Anatomy of a Table Saw

Large or small, the table saw is basically a motor and arbor assembly attached to a base cabinet or stand. The arbor may be mounted directly to the motor shaft, or connected to the motor by a belt and pulley. In general, better saws have more than one belt.

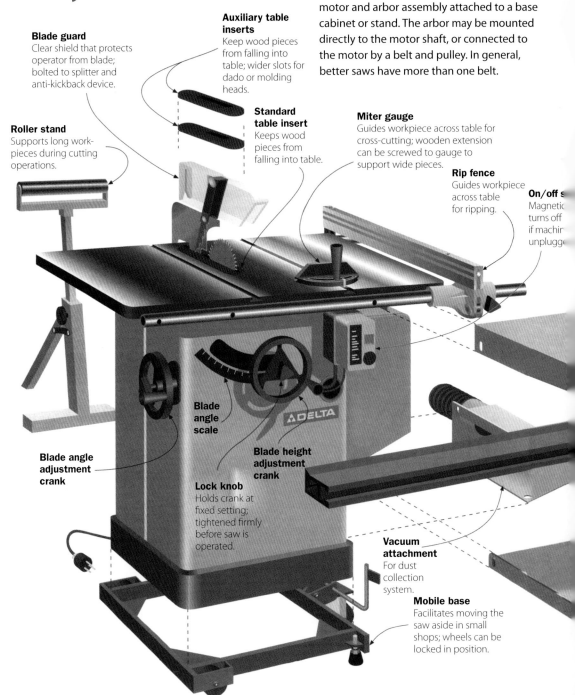

Blade guard
Clear shield that protects operator from blade; bolted to splitter and anti-kickback device.

Auxiliary table inserts
Keep wood pieces from falling into table; wider slots for dado or molding heads.

Roller stand
Supports long workpieces during cutting operations.

Standard table insert
Keeps wood pieces from falling into table.

Miter gauge
Guides workpiece across table for cross-cutting; wooden extension can be screwed to gauge to support wide pieces.

Rip fence
Guides workpiece across table for ripping.

On/off s
Magnetic turns off if machir unplugge

Blade angle scale

Blade height adjustment crank

Blade angle adjustment crank

Lock knob
Holds crank at fixed setting; tightened firmly before saw is operated.

Vacuum attachment
For dust collection system.

Mobile base
Facilitates moving the saw aside in small shops; wheels can be locked in position.

Woodworking Machines

Precise blade adjustments are made by means of two crank-type handwheels underneath the saw table. One wheel controls the blade's height above the saw table—from 0 to 3⅛ inches on a 10-inch saw. The other wheel adjusts the angle of the blade—from 90° to 45°.

The rip fence, which on most models slides along the front and rear guide bars to control rip cuts, can be locked anywhere along its track at the desired distance from the blade. Some fences feature measuring tapes attached to the front guide bar or even, in some cases, electronic readouts, although experienced woodworkers usually rely on a handheld measuring tape and a sample cut to check the width of a cut.

Shallow slots, milled into the saw table on each side of the blade, accept an adjustable miter gauge for guiding crosscuts. Quality saws have tables that are cast and then machined for flatness.

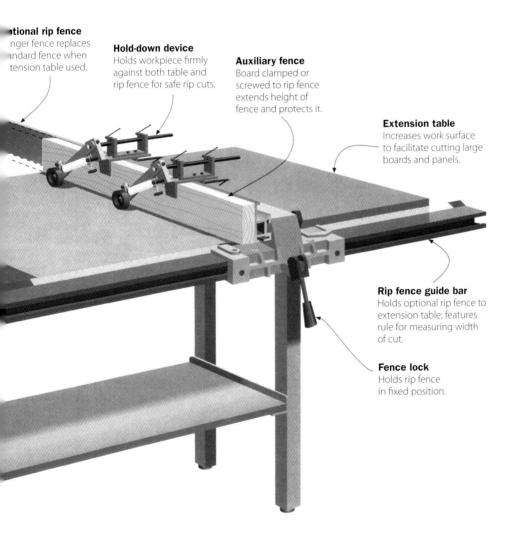

Optional rip fence
Longer fence replaces standard fence when extension table used.

Hold-down device
Holds workpiece firmly against both table and rip fence for safe rip cuts.

Auxiliary fence
Board clamped or screwed to rip fence extends height of fence and protects it.

Extension table
Increases work surface to facilitate cutting large boards and panels.

Rip fence guide bar
Holds optional rip fence to extension table; features rule for measuring width of cut.

Fence lock
Holds rip fence in fixed position.

Setting Up

Whether your table saw sits poised to make its first cut, or is a seasoned machine with a home full of furniture to its credit, it cannot cut with precision unless its adjustable parts are in proper alignment. A table saw with misaligned parts can result in any one of several frustrating problems, including excessive vibration, increased risk of kickback, blade damage, burn marks on workpieces, as well as inaccurate cuts. Even errors as little as $1/64$ inch can compromise the quality and strength of a piece of furniture.

The components of your table saw requiring the most attention are those that contact and guide the workpiece during cutting operations: the saw table, the blade, the miter gauge and the rip fence. Before putting a table saw through its paces on the cutting techniques described in this chapter, first set up the machine properly by checking and, if necessary, adjusting the alignment of its parts. For best results, unplug the saw, adjust the table insert setscrews to make the insert perfectly flush with the saw table, and crank the blade to its highest setting. Then follow the steps shown below in the sequence that they appear. There is little point in aligning the miter gauge with the saw blade, for example, if the blade itself has not been squared with the table.

To confirm that your table saw is properly tuned, make a few test cuts. A good way to ensure that your saw is cutting in precise, straight lines is to cut a squared board in two and flip one of the pieces over. Butt the two cut ends together. They should fit together without any gaps as perfectly as they did before the board was flipped.

Because the normal vibration from cutting can upset proper alignment, tune your table saw periodically; many woodworkers take the time to adjust their saws before starting each project.

Table Saws

The table saw is the cornerstone of many workshops, put to use in nearly every phase of every project. Because of its crucial role, your table saw must be consistently accurate and its parts square and true. The normal forces of routine use will eventually throw a table saw out of alignment. Even a new machine straight off the assembly line usually needs a certain amount of adjustment.

The table saw components that need to be checked and aligned are those that come in contact with the workpiece during the cut: the blade, table, miter gauge, and rip fence. If any of these parts is not aligned, you risk burn marks, tapered cuts, or kickback.

The simple tune-up procedures shown here will improve the performance of any table saw. Take the time to undertake them before starting a new project. For the sake of efficiency, follow the steps in the order they appear. You will only be able to align the miter gauge with the saw blade, for example, if the table has been squared with the blade. For safety, remember to unplug your saw before performing these checks and adjustments.

Most table saws feature worm gear and rack mechanisms connected to crank wheels to raise and tilt the arbor assembly and blade. These mechanisms can become caked with pitch and sawdust, preventing the saw from operating smoothly. In the photo above, compressed air is being used to clean the blade height mechanism.

Checking Table Alignment

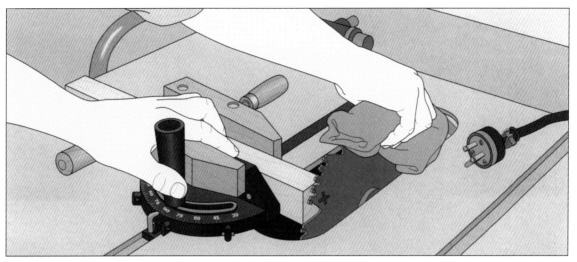

Checking the table alignment
The face of the miter gauge and the blade must be perfectly perpendicular. To check this, position the miter gauge at the front of the saw blade. Clamp a square wood block against the miter gauge with the end of the block butted against a saw blade tooth. Mark an X on the blade next to the tooth; this will enable you to check the same section of blade should you need to repeat the test after aligning the saw table. Slide the miter gauge and the block together toward the back of the table while rotating the blade by hand *(above)*. The block should remain butted against the tooth as the blade rotates from front to back. If a gap opens between the block and the tooth, or the block binds against the blade as it is rotated, you will need to align the table.

Aligning the saw table
Adjust the saw table following the owner's manual instructions. For the model shown, use a hex wrench to loosen the table bolts that secure the top to the saw stand *(left)*; the bolts are located under the table. Loosen all but one of the bolts and adjust the table position slightly; the bolt you leave tightened will act as a pivot, simplifying the alignment process. Repeat the blade test. Once the table is correctly aligned with the blade, tighten the table bolts.

Aligning the Table and Saw Blade

Checking table alignment
Position the miter gauge at the front of the saw blade. Hold or clamp a perfectly squared wood block against the miter gauge and butt the end of the block against a saw blade tooth *(right)*. Then slide the miter gauge and the block together toward the back of the table while rotating the blade by hand. The block should remain butted against the tooth as the blade rotates from front to back. If a gap opens between the block and the tooth, or the block binds against the blade as it is rotated, align the table following the owner's manual instructions.

Checking blade angle
Remove the table insert, then butt a combination square against the saw blade between two teeth as shown. The blade of the square should fit flush against the saw blade. If there is a gap between the two, rotate the blade angle adjustment crank until the saw blade rests flush against the square's blade.

Squaring the Miter Gauge

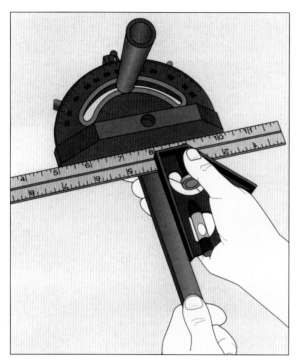

Aligning the miter gauge with the saw table
With the miter gauge out of the table slot, use a combination square to confirm that the face of the gauge is square with the edge of the gauge bar *(above, left)*. If it is not, use the adjustment handle on the gauge to square the two. Then place the miter gauge in its table slot and butt the square against the gauge *(above, right)*. The blade of the square should fit flush against the gauge. If there is a gap between the two, have the gauge machined square at a metalworking shop.

Aligning the miter gauge with the saw blade
Butt a carpenter's square against the miter gauge and the saw blade between two teeth. The square should fit flush against the gauge. If there is a gap between the two, loosen the adjustment handle on the gauge *(right)* and swivel the miter head to bring it flush against the square. Tighten the adjustment handle on the gauge.

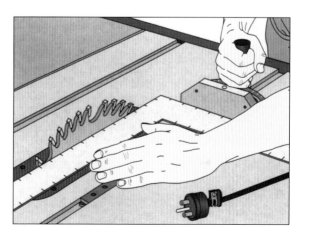

Testing the Table Saw for Square

Checking your adjustments
Test the accuracy of your table saw adjustments by crosscutting a couple of scrap boards. To check the blade-to-table alignment, mark an X on a board and cut it face down at your mark. Then turn the cutoff over and hold the cut ends together *(board A in the illustration at right)*. Any gap between the two ends represents twice the error in the table alignment; if necessary, repeat the test shown on page 62. To check the miter gauge adjustment, crosscut the second board, face down as well, flip one piece over, and butt the two pieces together on edge *(board B)*. Again, any gap represents twice the error in the adjustment. If necessary, square the miter gauge again *(page 63)*.

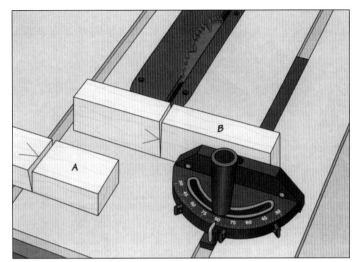

Aligning the Rip Fence

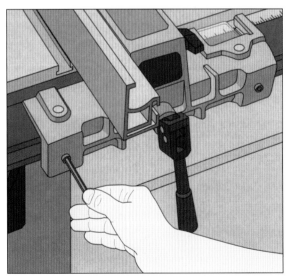

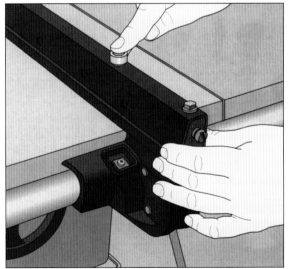

Adjusting the rip fence
Lock the rip fence in place alongside the miter slot. If the fence and the slot are not parallel, adjust the angle of the fence following the manufacturer's instructions. Some models feature adjustment bolts at the front of the table that you can loosen or tighten with a hex wrench to change the alignment *(above, left)*; others have fence adjustment bolts that you can loosen with a wrench *(above, right)*. For this model, adjust the fence parallel to the miter slot, then retighten the adjustment bolts.

Leveling the Table Insert

Adjusting the leveling screws

To set the table insert level with the saw table, place a square board across the insert and the table. Adjust the leveling screws at the corners of the insert with a hex wrench *(right)* until the insert is flush with the tabletop. You can also adjust the insert slightly below the table at the front and slightly above the table at the back; this will prevent the workpiece from catching or binding on the insert during the cut. If your saw's insert does not have leveling screws, file or shim the insert to make it lie flush with the table.

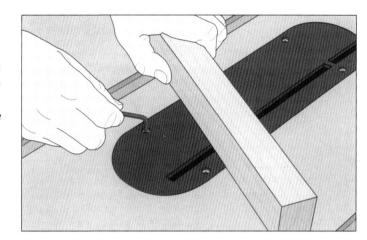

Adjusting the Height and Tilt Mechanisms

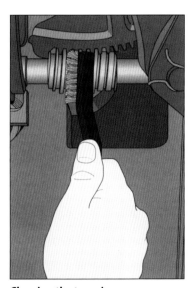

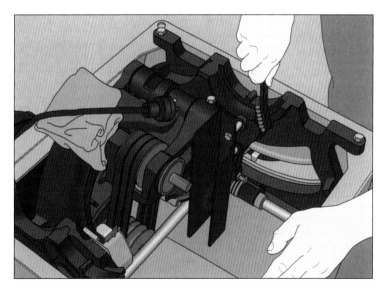

Cleaning the trunnions

If your table saw's blade sticks or moves sluggishly when you raise or tilt it, clean the height and tilt adjustment mechanisms inside the saw. Start by removing the tabletop following the manufacturer's instructions. Blow out the sawdust with compressed air, then clean the moving parts within the saw. Start with the blade height and tilt mechanisms *(above, left)*, using solvent and a brass-bristle brush to remove stubborn pitch and hardened sawdust deposits. Then scrub the machined ways on the front and rear trunnions *(above, right)*. Once all the parts are clean, lubricate all the moving parts with a graphite or silicon-based lubricant; oil and grease should be avoided as they tend to collect dust. Replace the tabletop and fine-tune the saw *(page 61)*.

Table Saw Blades

A table saw is only as good as the saw blade it turns. A dull or chipped blade can transform even the best of table saws into a poor or even dangerous tool. To protect blades from damage, avoid stacking them directly atop each other. Hang them individually on hooks or place cardboard between them. Replace a blade that is dull or cracked or has chipped teeth; more accidents are caused by dull blades than sharp ones.

Keep your saw blades clean. Wood resins can gum up a blade and hamper its ability to make a smooth cut. To clean sticky wood resin and pitch off a blade, soak it in turpentine, then scrub it with steel wool. Spray-on oven cleaner can be used to dissolve stubborn deposits.

Proper blade performance is as much a matter of using the right blade for the job as keeping it clean and in good condition. Whereas in the past there were relatively few saw blades to choose from, today's woodworker faces a wide array of options. As illustrated below, there are blades designed specifically for crosscutting or ripping, others to minimize kickback or produce thin kerfs, and blades for cutting specific types of wood. Regardless of type, all blades are installed on the saw and adjusted for cutting height and angle in the same way *(pages 61–62)*.

The most important advance in recent years has been the introduction of carbide-tipped blades. These have eclipsed traditional high-speed steel as the blade of choice. The advantage of carbide-tipped blades lies in their ability to keep a sharp edge far longer than their steel counterparts. Composed of grains of hard tungsten-carbon particles one-hundredth the thickness of a human hair, the carbide chunks are bonded with cobalt and brazed onto the blade with copper or silver. Carbide is extremely hard; the highest rating—C4—has a hardness value of 94 on a scale that rates diamond as 100.

While carbide-tipped blades can stay sharp for a hundred hours or more of use, they are more difficult—and therefore more expensive—to sharpen than high-speed steel blades. Still, most woodworkers believe the price is worth paying for the advantages they offer.

Rip Blade (Standard)
For cuts along the grain. Has deep gullets and relatively few, large teeth. The chisel-like cutting edges of the teeth make a fairly rough cut and produce large particles of sawdust and wood chips.

Crosscut Blade (Standard)
For cuts across the grain. Has more teeth than rip blade. The teeth make a smooth cut and produce fine sawdust.

Crosscut Blade (Anti-Kickback)
A variation of the standard crosscut blade. The projection between the teeth limits the size of the chips made with each bite; less aggressive bites prevent kickback.

Combination Blade
A general-purpose blade for ripping or crosscutting; does not make as smooth a cut as a rip or crosscut blade, but makes frequent blade changes unnecessary.

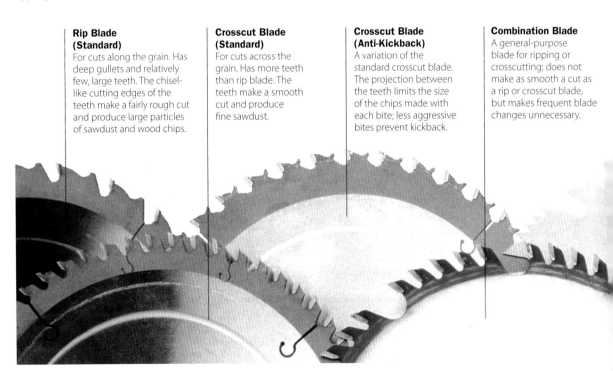

Woodworking Machines

Guide to Carbide—Tipped Blade Designs

Carbide-tipped saw blades feature four basic tooth designs. Each has its own particular advantages and applications. All blades have teeth that shear through the wood and gullets that clear away sawdust and wood chips from the kerf. Some blades also have rakers that cut out any material left in the kerf by the teeth. On some blades, the teeth are alternately beveled—that is, they shear stock alternately from one side, and then the other side of the cut.

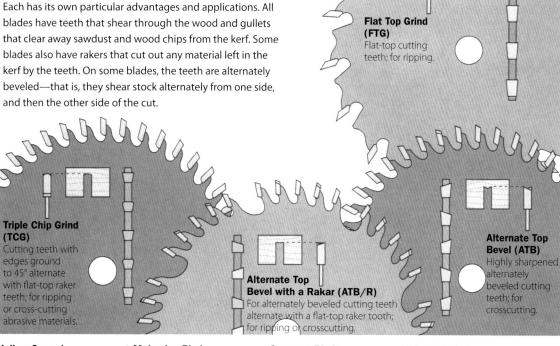

Flat Top Grind (FTG)
Flat-top cutting teeth; for ripping.

Triple Chip Grind (TCG)
Cutting teeth with edges ground to 45° alternate with flat-top raker teeth; for ripping or cross-cutting abrasive materials.

Alternate Top Bevel with a Rakar (ATB/R)
For alternately beveled cutting teeth alternate with a flat-top raker tooth; for ripping or crosscutting.

Alternate Top Bevel (ATB)
Highly sharpened alternately beveled cutting teeth; for crosscutting.

Hollow Ground Planer Blade (High-Speed Steel)
For very smooth crosscuts, rip cuts or angle cute. The body of the blade is thinner than the hub and teeth, which are not set, ensuring that the body does not bind in the saw kerf.

Melamine Blade
Has many small teeth designed to cut through the abrasive glue found in particleboard and other manufactured panels, resulting in a chip-free cut.

Crosscut Blade (Thin Rim)
A variation of the standard crosscut blade for fine finish cuts. Its thinner rim produces a narrower kerf, putting less strain on the saw motor.

Plywood Blade (High-Speed Steel)
Has many small teeth that make a smooth, splinter-free cut in plywood and wood veneers. The teeth are less efficient in highly abrasive manufactured panels such as particleboard.

Changing a Saw Blade

Removing the old blade
Working at the front of the table, wedge a piece of scrap wood under a blade tooth to prevent the blade from turning. Use the wrench supplied with the saw to loosen the arbor nut *(right)*. (Table saw arbors usually have reverse threads; the nut is loosened in a clockwise direction—not counterclockwise.) Finish loosening the nut by hand, making sure that it does not fall into the machine. Carefully lift the blade and washer off the arbor.

Installing the new blade
Slide the blade on the arbor with its teeth pointing in the direction of blade rotation (toward the front of the table). Insert the washer and nut and start tightening by hand. To finish tightening, grip the saw blade with a rag and use the wrench supplied with the saw *(left)*. Do not use a piece of wood as a wedge as this could result in overtightening the nut.

Setting the Blade Angle

Setting the proper cutting angle
To make an angle cut, remove the table insert and crank the blade to its highest setting. Use a protractor to set the desired cutting angle on a sliding bevel and butt the bevel against the blade between two teeth. Rotate the angle adjustment crank on the saw until the blade rests flush against the bevel *(right)*.

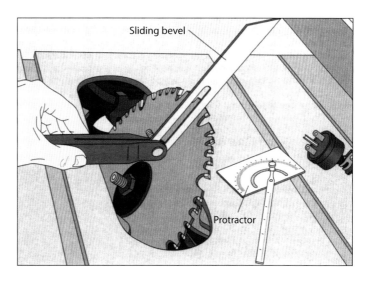

Setting the Blade Height

Cranking the blade to the proper height
A blade that is too high poses a safety risk; one that is too low will not cut properly. For most cutting operations, rotate the blade height adjustment crank until about ¼ inch of the blade is visible above the workpiece *(left)*. To set the blade at a specific height, use a tape measure or a commercially made gauge, which features a series of "steps" of ¼-inch increments; a similar gauge can be shop-built from scraps of ¼-inch plywood. The blade is at the correct height when it rubs the gauge as you rotate the blade by hand *(inset)*.

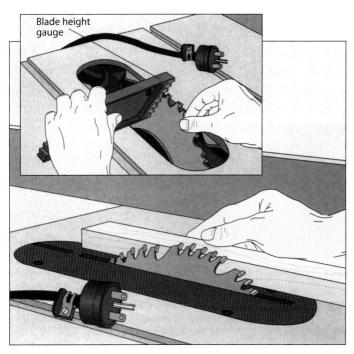

Ripping

Ripping has traditionally been defined as "cutting with the grain." But considering that some woods today—plywood and particleboard, for example—have no overall grain pattern, the definition needs some amending. A more appropriate description focuses on the table saw accessory used to make a rip cut. Whereas crosscutting is done using the miter gauge, ripping involves the rip fence. (Except for certain cuts that do not pass completely through the workpiece, such as a dado cut, the rip fence and miter gauge should never be used at the same time, or jamming and kickback can occur.)

Before ripping a workpiece, set the height of the saw blade *(page 71)*, then lock the rip fence in position for the width of cut. The most crucial safety concern when ripping is keeping your hands out of the blade's path. For protection, use accessories such as push sticks, featherboards and hold-down devices.

To use a hold-down device, it may first be necessary to screw a wood auxiliary fence to the rip fence. Auxiliary fences are ideal surfaces for clamping; many woodworkers make them a permanent fixture on their saws.

Ripping a Board

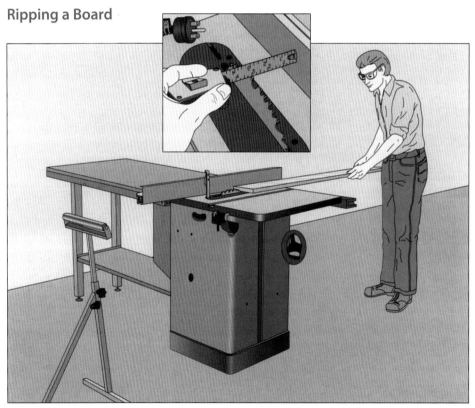

Starting the cut
Measure the distance to the edge of a tooth nearest the fence *(inset)*. Position the fence and set one end of the workpiece on the saw table close to the blade. Use your left hand to press the wood down on the table and flush with the fence; use your right hand to feed the wood into the blade *(above)*. Continue feeding the board into the blade at a steady rate until the trailing end of the board approaches the table. **(Caution: Blade guard removed for clarity.)**

Approaching the blade

Hook the thumb of your left hand over the edge of the table and rest your palm on the table, keeping the wood pressed down firmly on the table and up against the fence *(right)*. Continue feeding the board with your right hand until the trailing end of the board approaches the blade.

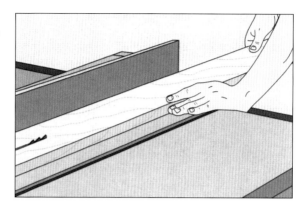

Passing the blade

Straddle the fence with your right hand *(left)*, making sure that neither hand is in line with the blade. If any finger comes within 3 inches of the blade, complete the cut using a push stick, a jig, or a hold-down device *(inset)* mounted on the rip fence. The rubber wheels of the hold-down device keep the workpiece firmly against the table; to prevent kickback, they also lock when pushed against the direction of the cut, keeping the board from shooting backward. If you are using a hold-down device, begin feeding the workpiece from the front of the table, then move to the back to pull the wood through. Otherwise, finish the cut from the front of the table.

Hold-down device

Finishing the cut

Keep pushing the board until the blade cuts through it completely. When the workpiece is clear of the blade, use your left hand to shift the waste piece to the left side of the table *(right)*. With your right hand, carefully lift the good piece and place it to the right of the rip fence before turning off the saw. Do not allow pieces of wood to pile up on the saw table.

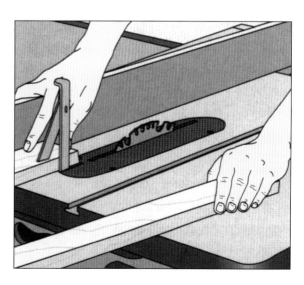

Ripping a Large Panel

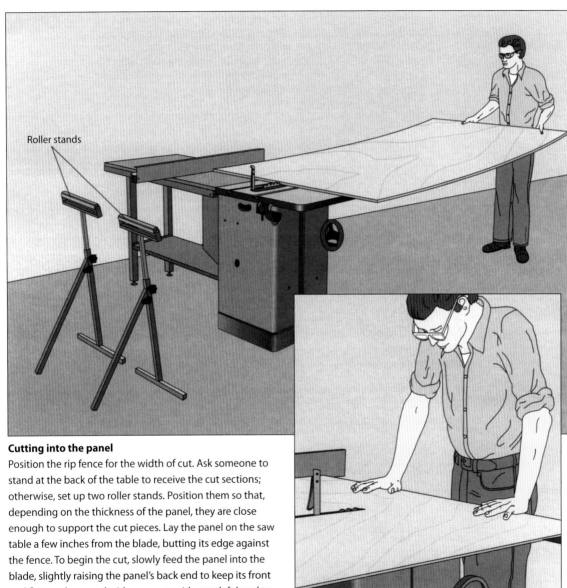

Cutting into the panel
Position the rip fence for the width of cut. Ask someone to stand at the back of the table to receive the cut sections; otherwise, set up two roller stands. Position them so that, depending on the thickness of the panel, they are close enough to support the cut pieces. Lay the panel on the saw table a few inches from the blade, butting its edge against the fence. To begin the cut, slowly feed the panel into the blade, slightly raising the panel's back end to keep its front end flat; apply enough side pressure with your left hand to keep the panel butted squarely against the fence *(above)*. Continue feeding the panel into the blade at a steady rate until its back end reaches the edge of the table. **(Caution: Blade guard removed for clarity.)**

Finishing the cut
Standing to the left of the saw blade, position your palms on the back end of the panel so that neither hand is in line with the blade. Press down on the panel with your palms *(above)* and push the trailing end of the panel toward the blade until the cut is completed.

Woodworking Machines

Ripping a Narrow Strip

Using a featherboard and push stick
Position the rip fence for the width of cut. Then butt the workpiece against the fence. To keep your hands away from the blade as it cuts the workpiece, use two accessories—a featherboard and a push stick. Clamp a featherboard to the saw table—the model shown is installed in the miter slot—so that its fingers hold the workpiece snugly against the fence. Use a push stick as shown to feed the workpiece into the blade. Continue cutting steadily until the blade nears the end of the cut. Support the waste piece with your left hand; to prevent your hand from being pulled back into the blade in case of kickback, curl your fingers around the edge of the table *(right)*. **(Caution: Blade guard removed for clarity.)**

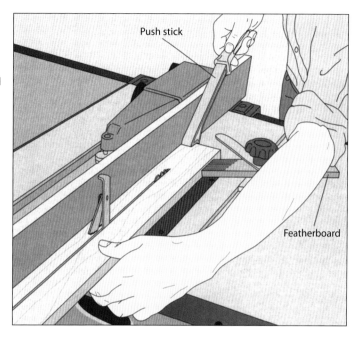

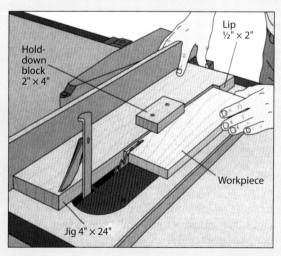

A Jig for Making Repeat Narrow Cuts

To rip several narrow strips to the same width, use the shopmade jig shown at left. For the jig, cut a board with a lip at one end. Screw a hold-down block to the jig, then butt the jig flush against the rip fence. Mark a cutting line on the workpiece, then seat it against the jig, flush with the lip. Position the rip fence so that the cutting line on the workpiece is aligned with the saw blade.

To make each cut, slide the jig and the workpiece as a unit across the table, feeding the workpiece into the blade *(left)*. (The first cut will trim the lip to the width of the cut.) Use your left hand to keep the workpiece flush against the jig. Remove the cut strip, reposition the workpiece in the jig, and repeat for identical strips. **(Caution: Blade guard removed for clarity.)**

Resawing Thick Stock

Setting up and starting the cut
To resaw a board, position the rip fence for the width of cut and attach a high auxiliary wood fence. Crank the blade below the table and place the workpiece over the table insert. To secure the workpiece, clamp one featherboard to the fence above the blade, and a second featherboard halfway between the blade and the front of the table. Rest the second featherboard on a wood scrap so that it supports the middle of the workpiece; clamp another board at a 90° angle to the featherboard for extra pressure, as shown. Remove the workpiece and set the blade height to a maximum of 1½ inches for softwood or 1 inch for hardwood. To start the cut, feed the workpiece into the blade *(right)*. Continue cutting at a steady rate until your fingers are about 3 inches from the blade.

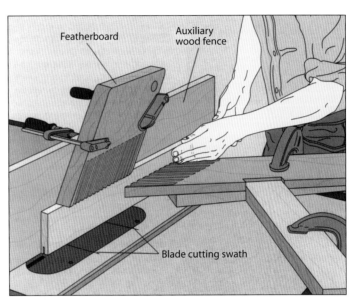

Completing the first pass
With the saw still running, move to the back of the table. Use one hand to press the workpiece flush against the rip fence *(above)* and the other hand to pull it past the blade. Flip the workpiece over and repeat the cutting procedures.

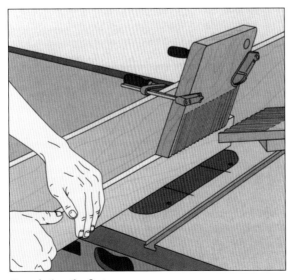

Finishing the cut
Raise the blade height and make another pass along each edge of the workpiece *(above)*. Make as many passes as necessary, raising the blade height after each pass, until the blade cuts through the workpiece completely.

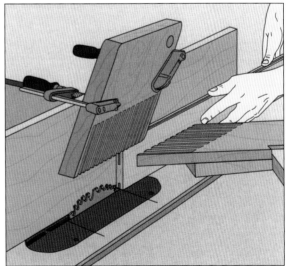

Taper Cuts

Using a commercial taper jig
To cut a workpiece so that one end is narrower than the other, make a taper cut. Hold the jig flush against the rip fence and pivot the hinged arm with the work stop until the taper scale indicates the cutting angle—in degrees or inches per foot. Mark a cutting line on the workpiece, then seat it against the work stop and hinged arm. Position the fence so that the cutting line on the workpiece is aligned with the saw blade. With the jig and workpiece clear of the blade, turn on the saw. Use your left hand to hold the workpiece against the jig and your right hand to slide the jig and workpiece as a unit across the table, feeding the workpiece into the blade *(right)*; ensure that neither hand is in line with the blade. Continue cutting at a steady rate until the blade cuts through the workpiece. **(Caution: Saw blade guard removed for clarity.)**

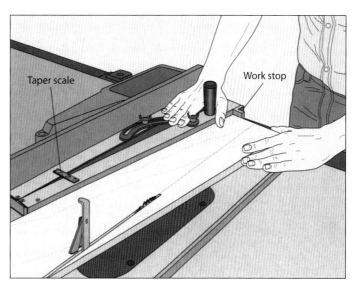

Using a shop-made jig
Build a jig exactly like the one shown on page 75 but without the handle. To position the workpiece for the taper cut, raise the saw blade to its highest setting. Butt one side of the jig base against the blade and position the rip fence flush against the other side of the base. Mark a cutting line on the workpiece, then place it on the base, aligning the line with the edge of the taper jig's base nearest the blade. Holding the workpiece securely, position the guide bar against it, with the lip snugly against the end of the workpiece. Screw the guide bar to the base and press the toggle clamps down to secure the workpiece to the jig base. Set the blade height. With the jig and workpiece clear of the blade, turn on the saw. With your left hand pressing the workpiece toward the rip fence, slide the jig and workpiece steadily across the table, making sure that neither hand is in line with the blade *(left)*. **(Caution: Blade guard removed for clarity.)**

Crosscutting

As cutting with the grain is synonymous with the use of the rip fence, so crosscutting is defined by the device used to make the cut: the miter gauge. The general technique for making a crosscut, as shown below, begins with correct hand placement to keep the workpiece both flush on the table and firmly against the miter gauge. The workpiece is fed into the blade at a steady rate. As with ripping, make sure that scrap pieces do not pile up on the table, and keep both hands out of line with the blade. Also, keep the rip fence well back from the blade to prevent any cut-off part of the workpiece from becoming trapped between the blade and fence and kicking back.

To reduce the amount of sanding you will need to do later, remember that the slower the feed, the smoother the cut, especially when the blade breaks through the workpiece at the end of the cut. Although a combination blade can be used for crosscutting, a crosscut blade will produce a finer cut.

When a longer workpiece is being cut, it is a good idea to attach an extension to the miter gauge to provide a more secure base. Miter gauges commonly have two screw holes for just such an addition—normally, a piece of hardwood 3 to 4 inches wide and about 2 feet long. Use the miter gauge extension in conjunction with a stop block to make repeat cuts *(page 80)*.

For wide panels or long boards, a shop-made crosscutting jig *(page 81)* is particularly helpful, and will ensure very accurate cuts. The jig can also be used for smaller pieces and provides a safe, convenient way to perform most crosscuts. Many experienced woodworkers consider it the single most indispensable accessory for crosscutting.

Squaring the Workpiece

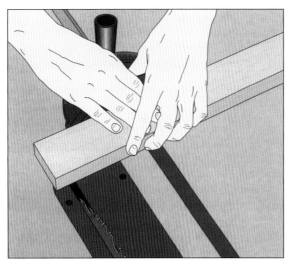

Making a crosscut
Before measuring or marking a workpiece for a crosscut, cut one end of it square. To avoid jamming the blade, align the workpiece with the blade so that it will trim ½ inch or so. With the thumbs of both hands hooked over the miter gauge, hold the workpiece firmly against the gauge *(above)* and push them together to feed the workpiece into the blade. **(Caution: Blade guard removed for clarity.)**

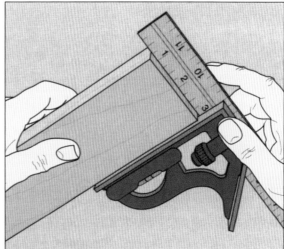

Checking for square
Use a combination square to confirm that the cut end of the workpiece forms a 90° angle with the edge. With the workpiece and square held up to the light, there should be no gap visible. Mark an X on the cut end to help you remember which end has been squared.

Repeat Cuts:
Using the Rip Fence as a Guide

Setting up the cut
Clamp a board to the rip fence as a stop block. To prevent jamming the workpiece between the stop and the blade—which could lead to kickback—position the stop far enough toward the front of the table so that the workpiece will clear the stop before reaching the blade. To line up the cut, hold the workpiece against the miter gauge and push the gauge and workpiece forward until the workpiece touches the saw blade. Slide the workpiece along the miter gauge until the cutting mark is aligned with the blade *(right)*.

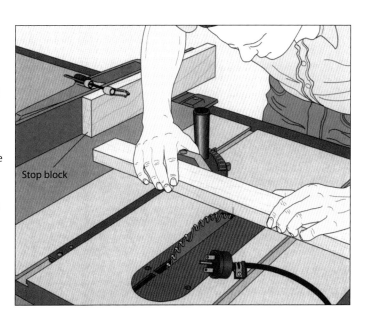

Positioning the rip fence
Holding the workpiece firmly against the miter gauge, pull both back from the blade and butt the stop block against the workpiece *(above)*. Lock the rip fence in position. Check to see that the workpiece does not contact the stop block when the workpiece reaches the blade.

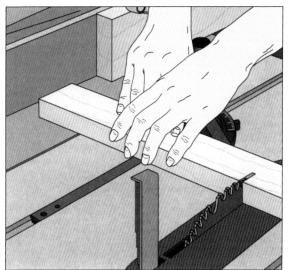

Making the cut
Set the end of the workpiece flush against the stop block. With the thumbs of both hands hooked over the miter gauge, hold the workpiece firmly against the gauge and push them together to feed the workpiece into the blade *(above)*. **(Caution: Blade guard removed for clarity.)**

Repeat Cuts: Using the Miter Gauge

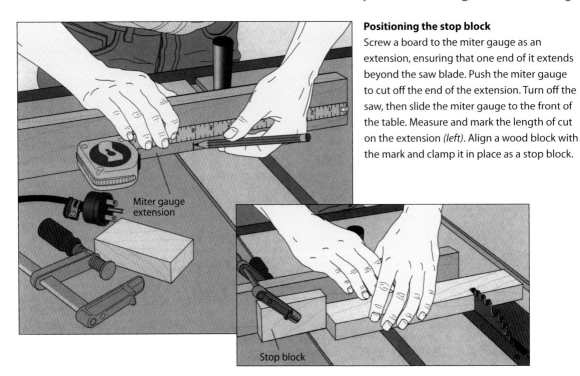

Positioning the stop block
Screw a board to the miter gauge as an extension, ensuring that one end of it extends beyond the saw blade. Push the miter gauge to cut off the end of the extension. Turn off the saw, then slide the miter gauge to the front of the table. Measure and mark the length of cut on the extension *(left)*. Align a wood block with the mark and clamp it in place as a stop block.

Making the cut
For each cut, butt the end of the workpiece against the stop block. With the thumbs of both hands hooked over the miter gauge, hold the workpiece firmly against the gauge and push them together, feeding the workpiece into the blade *(above)*. **(Caution: Blade guard removed for clarity.)**

Shop Tip

Hands-free "Off" switch
To turn off the saw when your hands are busy on the table, use a shop-made knee or foot lever. Cut a board equal in width to the switch box. The board should be long enough to reach with a foot or a knee when attached to the box *(right)*. Screw a hinge to one end of the board and position the hinge on top of the box. Mark the spot where the ON button touches the board. Cut a hole through the board at this mark. Attach the hinge to the box using glue, or remove the cover and drive in screws.

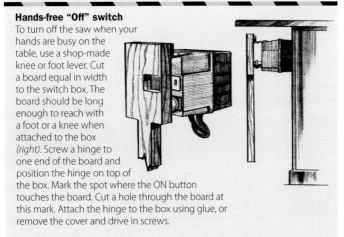

Woodworking Machines

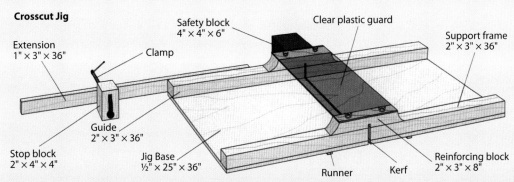

Crosscut Jig

- Extension 1" × 3" × 36"
- Safety block 4" × 4" × 6"
- Clamp
- Clear plastic guard
- Support frame 2" × 3" × 36"
- Guide 2" × 3" × 36"
- Stop block 2" × 4" × 4"
- Jig Base ½" × 25" × 36"
- Runner
- Kerf
- Reinforcing block 2" × 3" × 8"

For easy and accurate crosscuts—especially with long, wide or heavy workpieces—use a shop-built crosscut jig, custom-made for your table saw *(above)*. Refer to the illustration for suggested dimensions.

Cut two 25-inch-long hardwood runners the same width as your miter gauge slots. Bore clearance holes for screws into the undersides of the runners, 3 inches from each end. Place the runners in the slots, then slide them out to overhang the back end of the table by about 8 inches. Position the jig base squarely on the runners, its edge flush with their overhanging ends, then screw the runners to the base, countersinking the screws. Slide the runners and the base off the front end of the table and drive in the other two screws. Attach a support frame along the back edge of the jig. Glue a reinforcing block to the frame, centered between the runners. Then, with the runners in the miter gauge slots, make a cut through the support frame and three-quarters of the way across the base. Turn off the saw and lower the blade. Screw a guide to the front edge of the jig, ensuring that it is square with the saw kerf. Glue a safety block to the outside of the guide, centered on the kerf; also glue a reinforcing block on the guide, identical to the one on the support frame. Raise the saw blade and finish the cut, sawing completely through the guide but only slightly into the safety block.

For making repeat cuts to the same length, screw an extension to the guide and clamp a stop block to it. Use a clear plastic sheet that spans the saw kerf as a blade guard, fastening it to the reinforcing blocks with wing nuts.

To use the crosscut jig, fit the runners into the miter gauge slots. Slide the jig toward the back of the table until the blade enters the kerf. Hold the workpiece against the guide, slide the stop block to the desired position and clamp it in place, butting the end of the workpiece against the stop block. With the workpiece held firmly against the guide, slide the jig steadily across the table *(left)*, feeding the workpiece into the blade.

Crosscutting a Wide Panel

Reversing the miter gauge to start the cut
If a workpiece is wider than the distance between the front edge of the table and the saw blade, the miter gauge cannot be used to begin a crosscut in its usual position—in front of the blade. Instead, remove the gauge and insert it in the miter slot from the back of the table; for extra stability, screw a wooden extension to the gauge. To begin the cut, hold the extension with one hand while pressing the workpiece against it with the other hand. Feed the workpiece steadily into the blade until the trailing end of the workpiece reaches the front of the table. **(Caution: Blade guard removed for clarity.)**

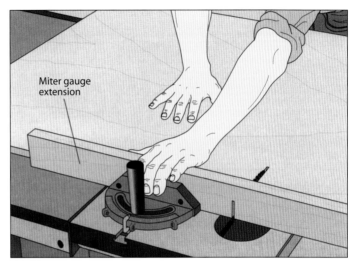

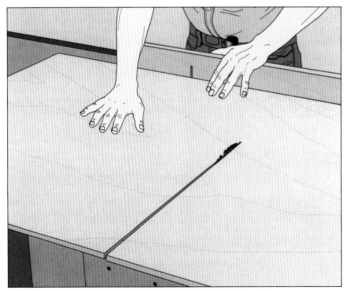

Finishing the cut
Turn off the saw when the blade is far enough through the workpiece to allow the miter gauge to return to its usual position, using a hands-free switch, if possible, so that both hands remain on the workpiece. Insert the miter gauge into its slot from the front of the table and complete the cut, holding the workpiece against the extension *(left)*.

Angle Cuts

One of the reasons the table saw is so versatile is that both the miter gauge and the blade can be angled, producing not only straight cuts but miter, bevel and compound cuts as well. Miters of between 30° and 90° are cut by angling the miter gauge. Saw blades can be tilted from 45° to 90°, producing bevel cuts. And by angling both the miter gauge and the saw blade, a woodworker can make a compound cut.

Whether crosscutting or ripping, the techniques used for angle cuts are similar to those used when the blade and gauge are at 90°. The difference is the result: With the blade at 90°, the woodworker ends up with a straight cut; with the blade angled, a bevel cut. The same applies to crosscutting, although with both activities extra care must be taken to keep hands away

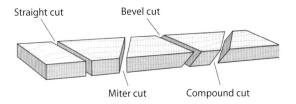

from the blade, which now cuts a wider swath above the table. When the blade is tilted, position the miter gauge or rip fence so that the blade angles away from it. This way the workpiece is pushed away from the blade rather than pulled toward it, reducing the chance that hands will stray into the blade. Gluing sandpaper to a miter gauge extension will also reduce the chance of a workpiece slipping during a cut.

Two Jigs for Making Multiple Angled Cuts

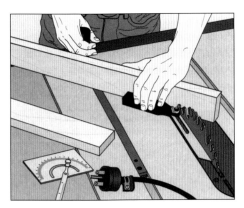

A simple setup for fast repeat cuts
Screw a wooden extension to the miter gauge, then use a sliding bevel to set the desired cutting angle of the gauge *(left)*. If you are making a compound cut, use the sliding bevel to set the blade angle. Push the miter gauge to cut off the end of the extension. Place the workpiece against the extension and line up the cutting mark with the blade. Clamp a stop to the extension at the opposite end of the workpiece. To make each cut, hold the workpiece firmly against the extension and, keeping both hands out of line with the saw blade, push the workpiece steadily into the blade.

Cutting miter joints
Build a crosscut jig without an extension or a safety block. Then, cut two 12-inch-long 1-by-4s and place them at 90° to each other in the middle of the jig, centered on its kerf. Turn the jig over and screw the 1-by-4s to the jig. To make a series of cuts, butt the workpiece against the left arm of the jig, align the cutting line on the workpiece with the saw blade and clamp a stop block to the arm at the end of the workpiece. Cut through the workpiece, holding it firmly against the arm and stop block *(right)*. Cut the mating piece of the joint the same way on the right arm of the jig. Use the stop blocks as guides for additional cuts to the same length.

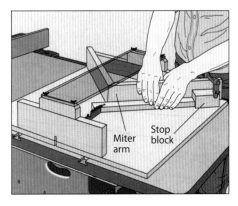

Band Saw

For ease of operation and wide-ranging utility, the band saw is hard to beat. It is the only woodworking machine capable of making both straight and contour cuts. In addition to crosscutting and ripping, it is well suited for cutting curves and circles, enabling the woodworker to produce anything from a dovetail joint to a cabriole leg.

Both rough and delicate work fall within its domain. Fitted with a ½-inch blade—the widest size available for most consumer-grade machines—a band saw can resaw 6-inch-thick lumber into two thinner pieces in a single pass. And with a ¹⁄₁₆-inch blade, a band saw can zigzag its way through a board at virtually any angle, even making 90° turns during a cut.

But do not let this tool's versatility intimidate you; the band saw is surprisingly easy to use. Many cuts can be made freehand by simply pivoting the workpiece around the blade. With the cutting techniques and shop-made jigs presented in this chapter, you will be able to turn out intricate curves, cut perfect circles and produce uniformly square-edged rip cuts and crosscuts.

One other advantage of the band saw over other woodworking machines is its relative safeness. Very little of the blade—usually only ⅛ inch—is ever exposed while it is running. And since the cutting action of the blade bears down on the workpiece, pushing it against the table instead of back toward the operator, kickback cannot occur. For this reason, the band saw is the tool of choice for ripping short or narrow stock.

Band saws are classified according to their throat width—that is, the distance between the blade and the vertical column, which supports the machine's upper wheel. Band saws for home workshops fall in the 10- to 14-inch range. Saws are also categorized according to their depth-of-cut capacity, which corresponds to the maximum gap between the table and the upper guide assembly. Although a 4- to 6-inch depth of cut is typical for consumer-grade saws, the band saw shown on pages 96–97 offers a height attachment that extends the vertical column to provide a 12-inch depth of cut. But even with a standard machine, you can take advantage of the band saw's unsurpassed depth-of-cut capacity by cutting identical patterns into several pieces of wood stacked one on top of another.

In choosing a band saw, look for one with a sturdy table that can tilt 45° in one direction and at least 10° in the other. In addition, consider spending a little more for a ¾-horsepower motor. For certain jobs, such as resawing a thick piece of stock, you will be glad to have the extra power.

This quarter-circle-cutting jig is an ideal time-saver for rounding corners for tabletops. The jig pivots around a fixed point, taking the guesswork out of cutting perfect arcs.

A ¼-inch band saw blade weaves its way along a curved cutting line, paring away a block of mahogany to form a graceful cabriole leg.

Angle Cuts

One of the reasons the table saw is so versatile is that both the miter gauge and the blade can be angled, producing not only straight cuts but miter, bevel and compound cuts as well. Miters of between 30° and 90° are cut by angling the miter gauge. Saw blades can be tilted from 45° to 90°, producing bevel cuts. And by angling both the miter gauge and the saw blade, a woodworker can make a compound cut.

Whether crosscutting or ripping, the techniques used for angle cuts are similar to those used when the blade and gauge are at 90°. The difference is the result: With the blade at 90°, the woodworker ends up with a straight cut; with the blade angled, a bevel cut. The same applies to crosscutting, although with both activities extra care must be taken to keep hands away

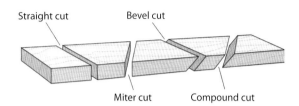

from the blade, which now cuts a wider swath above the table. When the blade is tilted, position the miter gauge or rip fence so that the blade angles away from it. This way the workpiece is pushed away from the blade rather than pulled toward it, reducing the chance that hands will stray into the blade. Gluing sandpaper to a miter gauge extension will also reduce the chance of a workpiece slipping during a cut.

Two Jigs for Making Multiple Angled Cuts

A simple setup for fast repeat cuts
Screw a wooden extension to the miter gauge, then use a sliding bevel to set the desired cutting angle of the gauge *(left)*. If you are making a compound cut, use the sliding bevel to set the blade angle. Push the miter gauge to cut off the end of the extension. Place the workpiece against the extension and line up the cutting mark with the blade. Clamp a stop to the extension at the opposite end of the workpiece. To make each cut, hold the workpiece firmly against the extension and, keeping both hands out of line with the saw blade, push the workpiece steadily into the blade.

Cutting miter joints
Build a crosscut jig without an extension or a safety block. Then, cut two 12-inch-long 1-by-4s and place them at 90° to each other in the middle of the jig, centered on its kerf. Turn the jig over and screw the 1-by-4s to the jig. To make a series of cuts, butt the workpiece against the left arm of the jig, align the cutting line on the workpiece with the saw blade and clamp a stop block to the arm at the end of the workpiece. Cut through the workpiece, holding it firmly against the arm and stop block *(right)*. Cut the mating piece of the joint the same way on the right arm of the jig. Use the stop blocks as guides for additional cuts to the same length.

Dado Cuts

Several woodworking joints call for channels to be cut into workpieces, allowing boards and panels to fit together tightly and solidly, but inconspicuously. Four of the most common types of channels are shown at right. They are distinguished from each other by their relationship to the wood grain and their location on a workpiece.

Each of these cuts can be made on a table saw with a standard blade by making repeated passes along the workpiece until the entire width of the channel is cut out. However, a table saw equipped with a dado head can cut a dado, groove or rabbet much more efficiently. There are several types of dado heads. The two most common are the adjustable wobble dado and the stacking dado shown below *(below)*.

The wobble dado is a single blade mounted on a hub that can be adjusted to provide varying widths of cut. Installed on the saw arbor much like a standard blade, the wobble dado literally wobbles as it spins. The greater the tilt—set by a dial on the blade—the wider the channel cut by the blade.

The stacking dado comprises a pair of outside blades that sandwich up to five inside chippers. The width of cut depends on how many chippers are mounted on the saw arbor along with the blades. Installing only the blades produces a ¼-inch cut. Inside chippers increase cutting width in $\frac{1}{16}$-, $\frac{1}{8}$- or ¼-inch increments up to $\frac{13}{16}$ inch—and up to 1 inch for models that include metal shims. Paper washers can be added to provide even finer width adjustment. For wider channels, adjust the dado head for the widest possible cut and make several passes.

Although adjustable wobble blades generally are less expensive and simpler to install than stacking models, most stacking dadoes provide better results: channels with more precise widths, flatter bottoms and cleaner edges with a minimum of tearout.

Rabbet: end-to-end cut at edge; either along or against the grain.

Groove: end-to-end cut along the grain.

Dado: end-to-end cut across the grain.

Stopped groove: cut along the grain that stops short of one or both ends.

Two Common Types of Dado Heads

Adjustable Wobble

Stacking Dado

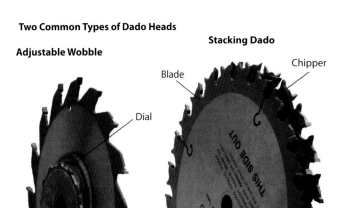

Blade · Dial · Chipper

From cutting grooves for shelves in a bookcase to making a rabbet to join two panels together, dado heads are an indispensable and versatile accessory for the table saw.

Installing a Dado Head

Installing blades and chippers
Remove the blade from the saw *(page 70)* and install a dado head following the manufacturer's instructions. For the carbide-tipped stacking dado shown, fit a blade on the arbor with the teeth pointing in the direction of blade rotation. To install a chipper, fit it on the arbor against the blade, with its teeth also pointing in the direction of blade rotation, and centered in gullets between two blade teeth. Fit additional chippers on the arbor the same way, offsetting their teeth from those of the chippers already in place. Then, fit the second blade on the arbor *(right)*, ensuring its teeth do not touch the teeth of the other blade or any chipper resting against it *(inset)*. Install the washer and tighten the nut on the arbor, keeping the blades and chippers in position, again making sure that the teeth of the chippers are not touching any blade teeth. If you cannot tighten the arbor nut all the way, remove the washer. Finally, install a dado table insert on the saw table.

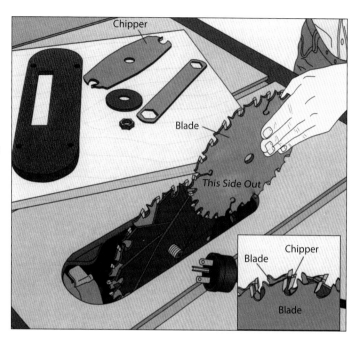

Making Dadoes and Grooves

Cutting a dado
Mark cutting lines for the width of the dado on the leading edge of the workpiece. Butt the cutting lines against the front of the dado head, then position the rip fence flush against the workpiece. Slide the workpiece to the front of the table and set it against the miter gauge—preferably with an extension screwed to it to provide extra stability. To make the cut, slide the miter gauge and the workpiece as a unit into the dado head *(left)*, keeping the workpiece firmly against the fence. (Since the dado head does not cut completely through the workpiece this is one exception to the general rule that the miter gauge and rip fence should never be used at the same time.) Continue feeding the workpiece at a steady rate until the cut is completed. **(Caution: Blade guard removed for clarity.)**

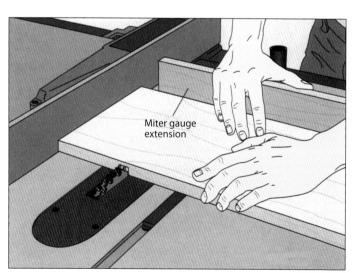

Cutting a groove

Mark cutting lines for the width of the groove on the leading edge of the workpiece. Butt the cutting lines up against the dado head, then position the rip fence flush against the workpiece. For narrow stock, use a featherboard and a push stick to keep your hands away from the dado head. Position your left hand at the front edge of the table to keep the trailing end of the workpiece flush against the fence. Feed the workpiece into the head *(right)* at a steady rate until the cut is completed. **(Caution: Blade guard removed for clarity.)**

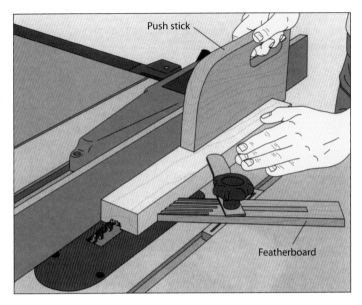

Cutting a rabbet

Install a dado head slightly wider than the rabbet desired, then crank it below the table. Screw a board to the rip fence as an auxiliary fence and mark the depth of the rabbet on it. Position the auxiliary fence directly over the dado head, ensuring that the metal fence is clear of the blade. Turn on the saw and slowly crank up the dado head until it cuts to the marked line, producing a relief cut in the auxiliary fence. Turn off the saw, then mark a cutting line for the inside edge of the rabbet on the workpiece. Butt the cutting line against the dado head, then position the rip fence flush against the workpiece. Clamp two featherboards as shown to hold the workpiece securely against the fence and saw blade; a wooden support arm provides extra stability. Turn on the saw, then feed the workpiece into the dado head *(left)* at a steady rate until the cut is completed; use a push stick, if necessary. **(Caution: Blade guard removed for clarity.)**

Making a Stopped Groove

Setting up the cut
To help you determine the position of the dado head when it is hidden by the workpiece during this cut, crank the dado head to the depth of the groove and use a china marker and a straightedge to mark the points where the head starts and stops cutting *(left)*. Then, mark two sets of cutting lines on the workpiece: one on its leading end for the width of the groove; one on its face for the length of the groove. Butt the cutting lines on the leading end of the workpiece against the front of the dado head, then position the rip fence flush against the workpiece.

Cutting into the workpiece
Turn on the saw and hold the workpiece just above the dado head, aligning the front cutting line on the workpiece with the dado head cutting mark on the table insert farthest from you. Holding the workpiece tightly against the fence, slowly lower it onto the head *(right)*, keeping both hands clear of the head. When the workpiece sits squarely on the table, feed it forward while pressing it against the fence.

Front cutting line

Finishing the cut
When your left hand comes to within 3 inches of the head, slide your hand along the top edge of the workpiece to the back of the table, hooking your fingers around the table's edge. Continue cutting at a steady rate until the back cutting line on the workpiece aligns with the dado head cutting mark closest to you. To complete the cut, lift the workpiece off the dado head with your right hand *(left)*, still steadying it against the fence with your left hand hooked around the edge of the table.

Moldings

A table saw is more than just a machine to cut wood. With the proper setup, a saw blade can serve as a milling device to cut cove moldings. And by replacing the saw blade with a molding head and different sets of cutters, a plain board can become an elaborate molding. Pieces of wood can be shaped separately and then glued together to form an impressive array of designs. The results range from crown moldings for a cabinet to decorative door and frame moldings—made at a fraction of the cost of their store-bought counterparts.

Molding cutters are sold in sets of three, which are installed in a molding head and then fastened onto the arbor. By passing the wood over the cutters repeatedly and raising the molding head slightly each time, a pattern is cut into the wood. The more passes, the deeper the inscription.

Like a dado head, a molding head requires its own table insert with a wide opening to accommodate the width of the cutters. A woodworker can make an insert for each set of cutters by placing a blank piece of wood in the table insert slot and slowly cranking up the molding head— much like making special inserts for saw blades.

Molding heads have a reputation for being dangerous and while there are always hazards involved when using a table saw, there is little risk when molding heads are used with proper care. A few points to keep in mind: Do not cut moldings on short lengths of wood; a piece should be at least 12 inches long. Also, do not cut moldings on narrow strips; cut the moldings on pieces at least 4 inches wide and then rip to width.

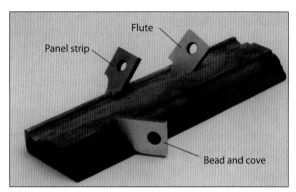

Milling baseboard with molding cutters
Three sets of cutters were used in combination to transform a piece of walnut into an elaborate baseboard molding at little cost *(above)*. More than 30 blade profiles are available; by using different cutters—also known as knives—on the same board, an innovative woodworker can mill an almost limitless range of designs.

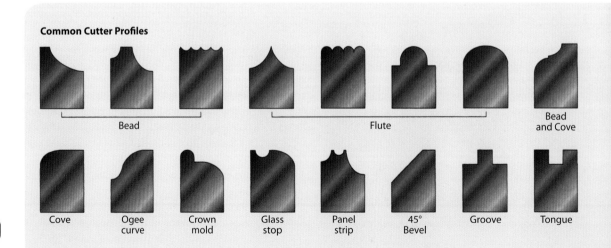

Common Cutter Profiles

Bead | Flute | Bead and Cove

Cove | Ogee curve | Crown mold | Glass stop | Panel strip | 45° Bevel | Groove | Tongue

Installing a Molding Head

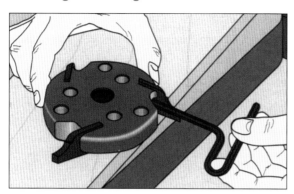

Mounting a molding head and cutters
Fit each of the three cutters partway into its slot in the molding head, ensuring that the cutter's beveled edge faces away from the setscrew hole. Install the setscrews into their holes, then use a hex wrench to tighten each screw until the cutters are seated firmly in their slots *(above)*. Install the molding head on the saw with the flat side of each cutter facing the direction of blade rotation. Grip the molding head with a rag to protect your hand and tighten the arbor nut counterclockwise using a wrench *(above)*. A washer is not necessary; the molding head is rigid enough without reinforcement. After the molding head is secured, install a molding-head table insert on the saw table. Rotate the molding head by hand to make sure that the cutters are true and that the unit does not rub against the insert.

Cutting a Molding

Featherboard

Supportboard

Setting up and making the first passes
Before cutting a molding, screw a board to the rip fence as an auxiliary fence. Position the auxiliary fence directly over the molding head, ensuring that the metal fence is clear of the cutters. Turn on the saw and crank up the molding head gradually to cut a notch in the auxiliary fence to allow for clearance of the cutters. Turn off the saw, then line up the cutting line on the end of the workpiece with the cutters and butt the rip fence against the workpiece. Crank the molding head to its lowest setting. To secure the workpiece, clamp one featherboard to the fence above the saw blade, and a second featherboard to the saw table. Clamp a support board at a 90° angle to the second featherboard, as shown. Remove the workpiece and crank the cutters to $\frac{1}{8}$ inch above the table; do not make a full-depth cut in one pass. Turn on the saw and use your right hand to slowly feed the workpiece toward the molding head; use your left hand to keep the workpiece against the rip fence. Finish the cut with a push stick. For a deeper cut, make as many passes as necessary *(left)*, raising the molding head $\frac{1}{8}$ inch at a time.

Making the final pass
After successive passes have produced the depth of cut desired, crank the molding head up very slightly and pass the workpiece through a final time at half the speed of previous passes *(left)*. By feeding the workpiece slowly, the final cut produces a smooth finish that requires minimal sanding.

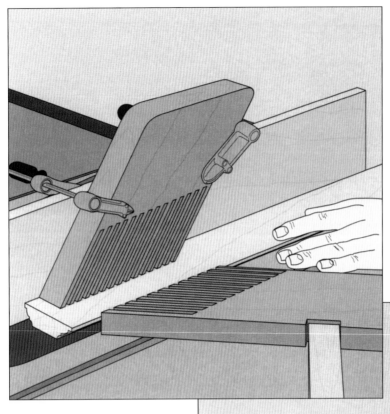

Separating the molding from the board
After the proper profile has been cut, separate the molding from the workpiece. Remove the molding head from the arbor and install a rip or combination blade. Feed the board through the blade, using a push stick to keep the workpiece firmly on the table *(right)*; use your left hand or a featherboard to press it flush against the rip fence.

Making a Box Joint

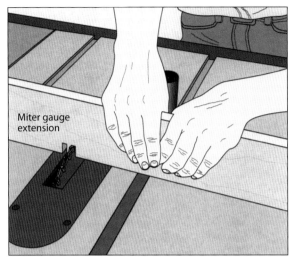

Setting up the jig
Cut the notches for a box joint one at a time using a dado head and jig. Clamp a board to the miter gauge as an extension. Crank the dado head to the desired height of the notches and feed the extension into the dado head to create a notch. Position the extension on the miter gauge so that the gap between the notch and the dado head is equal to the notch width, then screw the extension to the gauge. Feed the extension into the dado blade to cut a second notch *(above)*, checking that the gap between the notches equals the notch width. Fit and glue a hardwood key into the notch so that the key projects about an inch from the extension.

Cutting the notches in the first board
Butt one edge of the workpiece against the key, holding it flush against the miter gauge extension. To cut the notch, hook your thumbs around the gauge and slide the workpiece into the dado head *(above)*. Return the workpiece to the front of the table, fit the notch over the key and repeat the procedure. Continue cutting notches one after another until you reach the opposite edge of the workpiece.

Cutting the notches in the mating board
Fit the last notch you cut in the first board over the key, then butt one edge of the mating board against the first board, holding both flush against the miter gauge extension. To cut the first notch in the mating board, slide the two boards across the table *(right)*, then continue cutting notches in the mating board following the same procedure you used for the first board.

Making an Open Mortise-and-Tenon Joint

Cutting the tenon cheeks

Create a tenon by cutting the cheeks first, and then the shoulders. Install a commercial tenoning jig on the table following the manufacturer's instructions; the model shown slides in the miter slot. Mark cutting lines on the workpiece to outline the tenon, then clamp the workpiece to the jig. Crank the blade to the height of the tenon and position the jig so that one of the tenon cheek cutting lines is butted against the blade. Use the jig handle to slide the jig along the miter gauge slot; loosen the clamp handle to move it sideways. Slide the jig to the front of the table and turn on the saw, then use your right hand to push the jig forward, feeding the workpiece into the blade *(right)*. Continue cutting at a steady rate until the cut is completed. Pull the jig back to the front of the table and turn off the saw. Turn the workpiece around so that the remaining cutting line for the thickness of the tenon is butted against the blade. Cut along it the same way as you made the first cut.

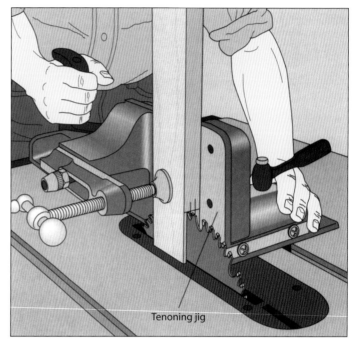

Sawing the tenon shoulders

Screw a board to the miter gauge as an extension. Then crank the blade to a height equal to the depth of the tenon against the extension, align one of the tenon shoulder cutting lines against the blade, then butt a stop block against the workpiece and clamp it in position. Slide the workpiece to the front of the table and turn on the saw. Hook your thumbs around the miter gauge to feed the workpiece into the blade and make the cut. Use a push stick to clear the waste piece off the table. Flip over the workpiece and butt it against the stop block, then cut the second shoulder *(left)*. **(Caution: Blade guard removed for clarity.)**

Making a Lap Joint

Cutting laps with a dado head
Mark cutting lines for the width of each lap on the leading edge of the workpiece. Butt one cutting line against the outside blade at the front of the dado head, then position the rip fence flush against the workpiece. Slide the workpiece to the front of the table and press it firmly against the fence and the miter gauge. To make the cut, slide the gauge and the workpiece as a unit into the dado head, keeping the workpiece flush against the fence. (This is another exception to the general rule that the miter gauge and rip fence should not be used at the same time.) Continue feeding the workpiece at a steady rate until the cut is made. Make successive passes *(right)*, cutting away the waste until the lap is completed. **(Caution: Blade guard removed for clarity.)**

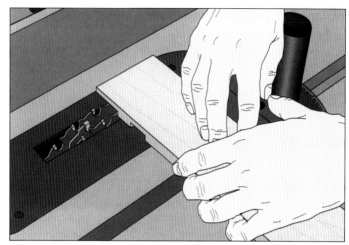

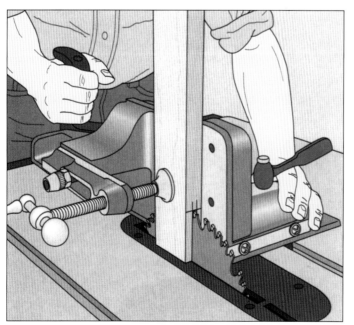

Cutting the mortise
Reinstall the tenoning jig on the table. Mark cutting lines on the workpiece to outline the mortise, then clamp the workpiece to the jig. Crank the blade to the depth of the mortise and position the jig so that one of the cutting lines is butted against the blade. Slide the jig to the front of the table, then turn on the saw and feed the workpiece into the blade. Pull the jig back and turn off the saw. Turn the workpiece over so that the remaining cutting line is butted against the blade and cut along it *(left)*. Make as many passes as necessary to remove waste between the two cuts. Test-fit the joint and deepen or widen the mortise, if necessary.

Band Saw

For ease of operation and wide-ranging utility, the band saw is hard to beat. It is the only woodworking machine capable of making both straight and contour cuts. In addition to crosscutting and ripping, it is well suited for cutting curves and circles, enabling the woodworker to produce anything from a dovetail joint to a cabriole leg.

Both rough and delicate work fall within its domain. Fitted with a ½-inch blade—the widest size available for most consumer-grade machines—a band saw can resaw 6-inch-thick lumber into two thinner pieces in a single pass. And with a 1/16-inch blade, a band saw can zigzag its way through a board at virtually any angle, even making 90° turns during a cut.

But do not let this tool's versatility intimidate you; the band saw is surprisingly easy to use. Many cuts can be made freehand by simply pivoting the workpiece around the blade. With the cutting techniques and shop-made jigs presented in this chapter, you will be able to turn out intricate curves, cut perfect circles and produce uniformly square-edged rip cuts and crosscuts.

One other advantage of the band saw over other woodworking machines is its relative safeness. Very little of the blade—usually only ⅛ inch—is ever exposed while it is running. And since the cutting action of the blade bears down on the workpiece, pushing it against the table instead of back toward the operator, kickback cannot occur. For this reason, the band saw is the tool of choice for ripping short or narrow stock.

Band saws are classified according to their throat width—that is, the distance between the blade and the vertical column, which supports the machine's upper wheel. Band saws for home workshops fall in the 10- to 14-inch range. Saws are also categorized according to their depth-of-cut capacity, which corresponds to the maximum gap between the table and the upper guide assembly. Although a 4- to 6-inch depth of cut is typical for consumer-grade saws, the band saw shown on pages 96–97 offers a height attachment that extends the vertical column to provide a 12-inch depth of cut. But even with a standard machine, you can take advantage of the band saw's unsurpassed depth-of-cut capacity by cutting identical patterns into several pieces of wood stacked one on top of another.

In choosing a band saw, look for one with a sturdy table that can tilt 45° in one direction and at least 10° in the other. In addition, consider spending a little more for a ¾-horsepower motor. For certain jobs, such as resawing a thick piece of stock, you will be glad to have the extra power.

This quarter-circle-cutting jig is an ideal time-saver for rounding corners for tabletops. The jig pivots around a fixed point, taking the guesswork out of cutting perfect arcs.

A ¼-inch band saw blade weaves its way along a curved cutting line, paring away a block of mahogany to form a graceful cabriole leg.

Woodworking Machines

Band Saw

Back to **Basics**

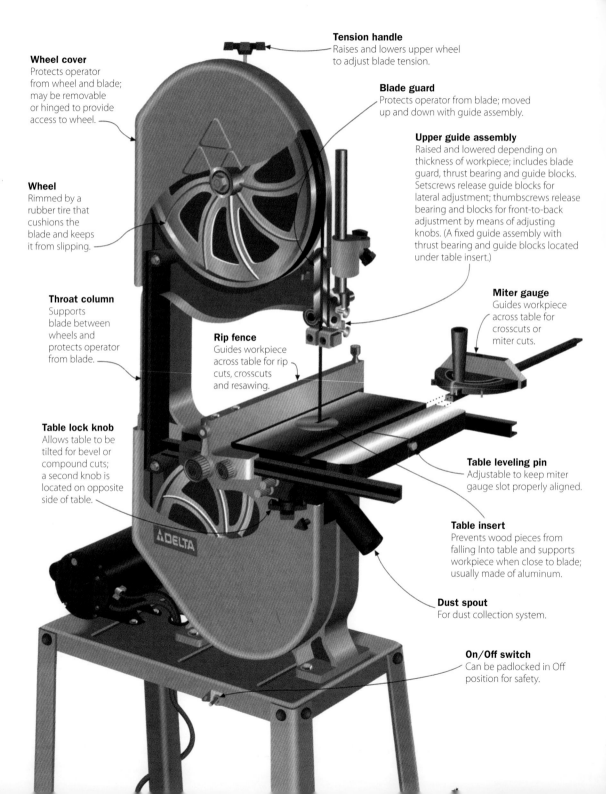

Woodworking Machines

Anatomy of a Band Saw

As the name suggests, a band saw blade is a continuous steel band. Varying in length from roughly 72 inches to 104 inches depending on the size of the machine, the blade runs around rubber-rimmed wheels and passes through an opening in the saw table. One of the wheels—typically the lower one—is the drive wheel, which is turned by a motor. The blade is not fastened to the wheels but is held in place by tension and turns through its elliptical path at roughly 3,000 feet per minute—the average cutting speed for a 14-inch saw.

The blade is kept taut by means of a tension handle, which raises and lowers the upper wheel. A tilt knob that cants the upper wheel is used to keep the blade centered on the wheels. The blade is kept steady on its path by thrust bearings located behind the blade above and below the table, and by guide blocks, which prevent lateral movement. Although some cuts can be made freehand, a rip fence and miter gauge are available with many models to guide workpieces across the table.

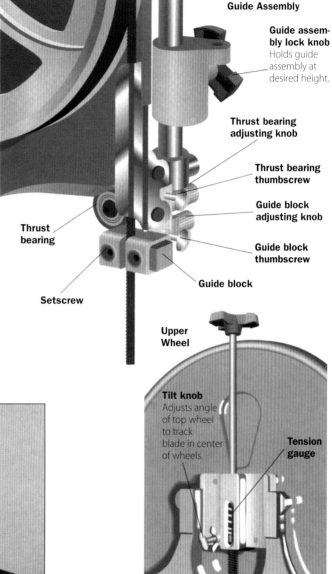

The three-wheel band saw's wide throat capacity—typically 20 inches, rather than the 10 to 14 inches available on most two-wheel models—makes it more convenient for working with particularly large workpieces.

Setting Up

The band saw has a reputation among some woodworkers as a relatively imprecise cutting tool. And yet band saws are routinely used in industry to cut very hard materials such as metal to very close tolerances. The fact remains, however, that the tool can only be made to cut straight edges and precise curves if it is kept finely tuned.

The ideal is for the blade to cut squarely into the workpiece, producing a smooth, accurate result. But the peculiarities of band saw geometry can make this ideal difficult to achieve. After bending around the machine's wheels at 35 miles per hour, a section of the blade must straighten out by the time it reaches the saw table a split second later.

For this to happen, the adjustable parts of the saw must be kept in proper alignment so that the blade runs smoothly and square to the table. Particular attention should be paid to the wheels, the guide assembly and the saw table itself.

To tune your band saw, unplug it, install and tension the blade you plan to use *(page 101)* then follow the set-up steps detailed on the following pages. Take the time to do it right. Adjusting the band saw may be more time-consuming than learning how to operate the tool. But the advantages of a well-tuned machine will be noticeable not only in the quality of the results but in the longevity of your blades and of the band saw itself. Misaligned wheels or poorly adjusted guide blocks can lead to premature blade wear or breakage.

Installing nonmetallic guide blocks on a band saw can reduce wear and tear appreciably *(page 102)*, but there is no substitute for getting around the need to check thrust bearings, guide blocks and wheels for proper alignment.

Checking the Wheels

Checking the wheel bearings
Open one wheel cover, grasp the wheel at the sides, and rock it back and forth *(left)*. Repeat while holding the wheel at the top and bottom. If there is play in the wheel or you hear a clunking noise, remove the wheel and replace the bearing. Then repeat the test for the other wheel.

Woodworking Machines

Testing for out-of-round wheels
Start with the upper wheel. Bracing a stick against the upper guide assembly, hold the end of the stick about 1/16 inch away from the wheel's tire. Then spin the wheel by hand *(right)*. If the wheel or tire is out of round, the gap between the stick and the wheel will fluctuate; the wheel may even hit the stick. If the discrepancy exceeds 1/32 inch, remedy the problem. Repeat the test for the lower wheel.

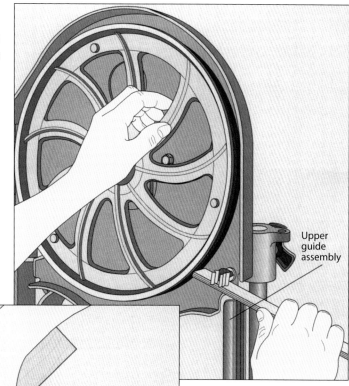

Upper guide assembly

Fixing an out-of-round wheel
Start by determining whether the tire or the wheel itself is the problem. Try stretching the tire into shape with a screwdriver, then repeat the earlier test. If the wheel is still out of round, use a sanding block to sand the tire; this may compensate for unevenness in the tire. For the lower wheel, turn on the saw and hold the sanding block against the spinning tire *(left)*. For the upper wheel, leave the saw unplugged and rotate the wheel by hand. Test the wheel again. If the problem persists, the wheel itself is out of round. Have it trued at a machinist's shop.

Aligning the Wheels

Checking wheel alignment
To make certain that the wheels are parallel to each other and in the same vertical plane, loosen the table lock knobs and tilt the table out of the way. Open both wheel covers and hold a long straightedge against the wheel rims as shown. The straightedge should rest flush against the top and bottom of each wheel. If the wheels are out of alignment, try to bring the top wheel to a vertical position by means of the tilt knob. If the straightedge still will not rest flush, you will have to adjust the position of the upper wheel.

Shifting the upper wheel
Move the upper wheel in or out on its axle following the instructions in your owner's manual. On the model shown, you must first remove the blade *(page 106)* and the wheel. Then shift the wheel by either adding or removing one or more washer *(left)*. Reinstall the wheel and tighten the axle nut. Install the blade and recheck wheel alignment.

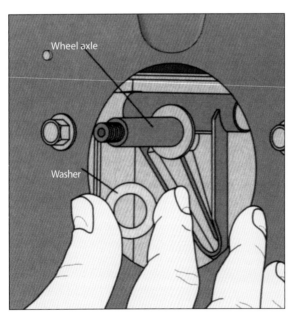

Tensioning and Tracking a Blade

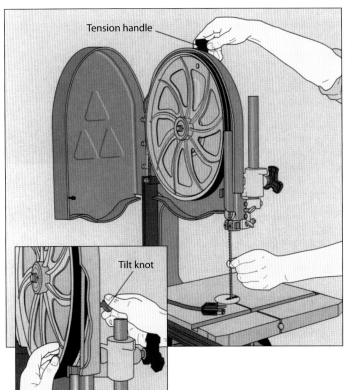

Tensioning a blade
Turn the tension handle clockwise with one hand to raise the top wheel and increase tension on the blade; deflect the blade from side to side with the other hand to gauge the tension. Spin the upper wheel by hand and gauge the tension at several points along the blade. Increase the tension *(left)* until the blade deflects about ¼ inch to either side of the vertical position. Avoid overtensioning a blade; this can lead to premature blade wear and breakage. Undertensioning a blade will allow it to wander back and forth and side to side as it cuts.

Tracking a blade
Lower the upper guide assembly, then spin the upper wheel by hand to check whether the blade is tracking in the center of the wheel. If it is not, loosen the tilt knob lock screw. Then, spin the wheel with one hand while turning the tilt knob with the other hand *(above)* to angle the wheel until the blade tracks in the center. To check the tracking, close the wheel covers and turn on the saw, then turn it off; adjust the tracking, if necessary. Set the thrust bearings and guide blocks *(page 102)*.

Shop Tip

Rounding a blade
To help prevent a new band saw blade from binding in the kerf of a curved cut, use a silicon-carbide stone without oil to round its back edge. Glue the stone onto a shop-made handle. Tension and track the blade, then turn on the saw. Wearing safety goggles, hold the stone against the back of the blade and slowly pivot the stone. Turn off the saw after a few minutes. In addition to rounding the blade, the stone will smooth any bumps where the blade ends are welded together.

Adjusting the Guide Assemblies

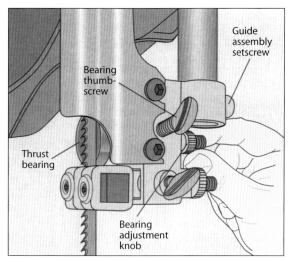

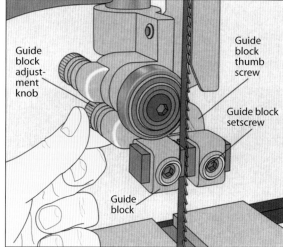

Setting the thrust bearings
Set the upper guide assembly then check by eye that the upper thrust bearing is square to the blade. If not, loosen the guide assembly setscrew, adjust the assembly so that the bearing is square to the blade, and tighten the setscrew. Then, loosen the bearing thumbscrew and turn the adjustment knob until the bearing just touches the blade. Back the bearing off slightly *(above)* and tighten the thumbscrew. (The lower thrust bearing, which is located directly under the table insert, is adjusted the same way.) To check the setting, spin the upper wheel by hand. If the blade makes either bearing spin, back the bearing off slightly and recheck.

Setting the guide blocks
To set the upper guide blocks, loosen the guide block setscrews and pinch the blocks together using your thumb and index finger until they almost touch the blade. Alternatively, use a slip of paper to set the space between the blocks and the blade. Tighten the setscrews. Next, loosen the thumbscrew and turn the adjustment knob until the front edges of the guide blocks are just behind the blade gullets *(above)*. Tighten the thumbscrew. Set the lower guide blocks the same way.

Heat-resistant guide blocks
Designed to replace the metal guide blocks supplied with most saws, nonmetallic blocks are made from a graphite-impregnated resin that contains a dry lubricant. Because they build up less heat than conventional guide blocks, the nonmetallic variety last longer; they can also be set closer to the blade, promoting more accurate and controlled cuts. In addition, contact between the blade and nonmetallic blocks does not dull the blade, as is common with metal blocks. To install, unscrew the guide block setscrews, remove the old blocks and replace with the new blocks; tighten the setscrews.

Squaring the Table and Blade

Aligning the table
To ensure that the miter gauge slot is properly aligned on both sides of the table slot, set the miter gauge in its slot and slide the gauge back and forth across the table. The gauge should slide freely with only moderate pressure. If the gauge binds, use locking pliers to remove the leveling pin. Then, insert the pin into its hole and use a ball-peen hammer to tap the pin deeper *(right)* until the miter gauge slides freely.

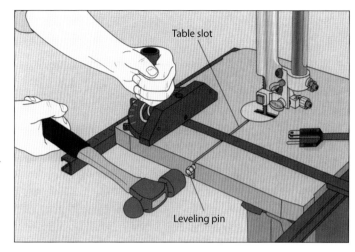

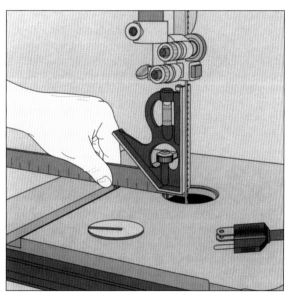

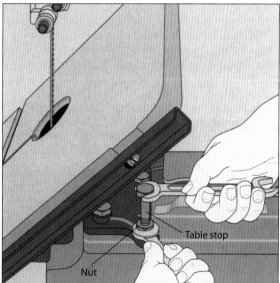

Checking the table angle
With the table in the horizontal position, remove the table insert, then butt a combination square against the saw blade as shown. The square should fit flush against the saw blade. If there is a gap between the two, loosen the two table lock knobs and make sure the table is seated properly on the table stop under the table. Tighten the lock knobs. If the gap remains, adjust the table stop.

Adjusting the table stop
Tilt the table out of the way, then use two wrenches as shown to adjust the table stop. Use the lower wrench to hold the nut stationary and the upper wrench to turn the table stop: clockwise to lower it; counterclockwise to raise it. Recheck the table angle.

Safety

Compared to the table saw or radial arm saw, the band saw seems like a relatively safe machine. There is no aggressive whine of a 1½- or 3-horse-power motor turning a 10-inch saw blade; instead, the band saw produces a quiet hum that some woodworkers liken to the sound of a sewing machine. And with its blade guard properly set, no more than ⅛ inch of the blade is exposed above the table.

Still, it is impossible to be too careful with any woodworking machine and the band saw is no exception. Band saw blades occasionally break, and when they do they tend to fly to the right of where the operator normally stands. Therefore, it is wise to stand slightly to the throat column side of the blade whenever possible. If a blade snaps, turn off the saw and do not open the wheel covers to install a new blade until the wheels have stopped completely.

Although the blade guard adequately covers the blade above the table, there is no guard at the level of the table or underneath it. As a result, you need to keep your hands out of the hole covered by the table insert and refrain from reaching under the table to clear debris from the blade before the blade has come to a stop.

Most of the accidents that occur with the band saw are a result of excessive feed pressure and poor hand position. Feed a workpiece steadily into the blade, but with a minimal amount of pressure, otherwise the blade may jam and break. For most cuts, feed the workpiece with one hand, using the other hand to guide it. Keep your fingers out of line with the blade. Hook the fingers of the feed hand around an edge of the workpiece to prevent them from slipping into the blade as your hand nears the cutting area.

Cutting Safely with the Band Saw

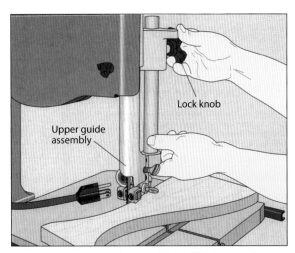

Setting the upper guide assembly and blade guard
Before turning on the saw to begin a cut, set the upper guide assembly ⅛ inch above the workpiece. Use one hand to hold the guide assembly in position and the other hand to tighten the guide assembly lock knob *(above)*. Alternatively, use the workpiece to lever the guide assembly up slightly, then tighten the lock knob. Setting the guide assembly as close to the workpiece as possible not only protects you from the blade when the saw is running; it also supports the blade as it cuts, minimizing excessive blade deflection.

Band Saw Safety Tips

- Except when changing a blade, always keep the wheel covers closed.

- Make sure that saw blades are sharp, clean and undamaged. Disconnect the saw before changing a blade.

- Stand slightly to the left of the blade when cutting at the front of the band saw table. Do not stand, or allow anyone else to stand, to the right of the blade. This is the direction in which the blade will fly if it breaks.

- Do not cut until the blade is turning at full speed.

- Keep your hands away from the blade when the saw is on. Use a push stick or a jig to cut small or narrow pieces.

- Avoid making turns that are too tight for the blade you are using. This can break the blade.

- Cut with the blade guard no more than ⅛ inch above the workpiece.

- Before backing out of a cut, turn off the saw.

- Always unplug the saw before doing any work on it.

Band Saw Blades

Lumber mill band saws regularly use blades as wide as 12 inches to cut logs into boards. Blades for consumer-grade saws are much smaller—generally ranging from 1/16 to 1/2 inch wide. But even within this relatively narrow spectrum, choosing the best blade for the job is not always straightforward. There is no single all-purpose combination blade in band sawing, nor any blade specifically designed for ripping or crosscutting. However, a woodworker should keep three basic variables in mind: tooth design, blade width and blade set.

As illustrated at right, band saw blades for cutting wood are available in three basic tooth designs; each design does something better than the others. The chart below shows the importance of selecting a blade of appropriate width for cutting curves. In general, narrow blades are used for cuts with intricate curves, while wide blades are ideal for resawing thick stock.

Blade set refers to how much the blade teeth are angled to the side, making a saw cut—or kerf—that is wider than the blade. This reduces the chance of the blade binding in a cut. A blade with minimal set, called a light set blade, produces a smooth cut and a narrow kerf, but is also more prone to binding, which limits its ability to cut a tight curve. A heavy set blade—one with greater set—cuts faster than a light set blade, and is less likely to bind due to its wider kerf. However, a heavy set blade leaves more visible corrugated marks in the cut edge of a workpiece, an effect called "washboarding."

There are enough stresses on a band saw blade under the best of circumstances without adding to them by improper operation of the machine. Some of the many avoidable causes of blade breakage include forcing a blade around a curve that is too tight for its width, improper adjustment of the blade guides, excessive feed speed or pressure, dull blade teeth, excessive blade tension, insufficient tooth set and running the blade for extended periods without cutting. Tension and track a blade *(page 101)* immediately after you install it. Incorrect tension can shorten the life of a blade.

The typical band saw blade has a loop length of several feet. To reduce the amount of storage space, fold the blade into three loops. Clean a band saw blade regularly to keep it from gumming up with resins and pitch. Use a wire or stiff-bristled brush dipped in solvent such as turpentine, oven cleaner or an ammonia-based cleaner. Before storing a blade or for removing rust, wipe the blade with an oily rag. For rust, use steel wool.

Blade Types

Standard Blade
For straight cuts across the grain or diagonal to the grain. Ideal for intricate curves or cuts when the orientation of the blade to the grain changes during the cut.

Skip-tooth Blade
So called because every other tooth is missing. For long, gentle curves with the grain. Cuts faster, but more roughly, than a standard blade. A 1/4-inch skip-tooth blade with 4 to 6 teeth per inch is a good all-purpose blade.

Hook-tooth Blade
For straight cuts and curves with the grain; the best blade for ripping or resawing.

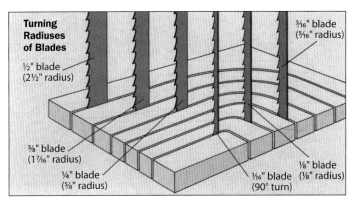

Turning Radiuses of Blades
- 1/2" blade (2 1/2" radius)
- 3/8" blade (1 7/16" radius)
- 1/4" blade (5/8" radius)
- 1/16" blade (90° turn)
- 1/8" blade (1/8" radius)
- 3/16" blade (5/16" radius)

When choosing a band saw blade for a contour cut, consider the tightest curve that the blade will turn. Use the chart at left as a rough guide. In general, the narrower the blade, the tighter the curve, given the same blade set. But because wider blades resist unwanted deflection, a narrow blade is not always the best choice for a curved cut. A good rule of thumb is to use the widest blade for the tightest curve required. The limitations on a blade's turning capacity cannot be ignored. Forcing a blade around a corner that is too tight will cause it to bind in the kerf, twist and, ultimately, snap.

Changing a Saw Blade

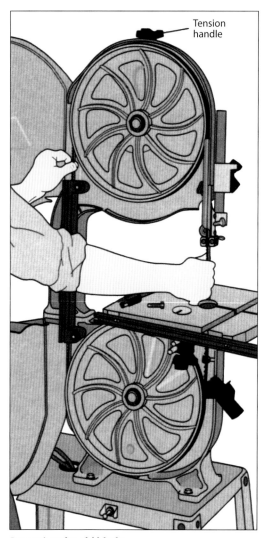

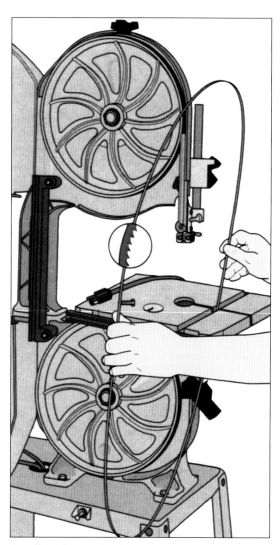

Removing the old blade
Raise the upper guide assembly to its highest setting and lock it in place *(page 101)*. Back the thrust bearings and guide blocks away from the blade *(page 102)*. Remove the table insert and use locking pliers to remove the table leveling pin. Turn the tension handle counterclockwise to release the blade tension, then open the wheel covers. Wearing safety goggles, carefully slide the blade out of the guide assemblies *(left)*, then slip it off the wheels and guide it through the table slot.

Installing the new blade
If the blade is coiled, uncoil it carefully. Band saw blades store a considerable amount of spring. Wearing safety goggles and gloves, hold the blade at arm's length in one hand and turn your face away as the blade uncoils. Guide the blade through the table slot as shown, holding it with the teeth facing you and pointing down. Slip the blade between the guide blocks and in the throat column slot, then center it on the wheels. Install the leveling pin and table insert. Tension and track the blade *(page 101)*.

Cutting Curves

Much of the curved wood that graces well-made furniture is cut on the band saw, which can produce virtually any contour. As shown in the pages that follow, you cut curves in a variety of ways: by sawing freehand along a cutting line, by making use of a pattern *(page 109)* or by relying on shop-built jigs.

Whatever the shape of the curve, the biggest challenge in contour-cutting is avoiding dead ends, where the workpiece hits the throat column before the end of a cut. When this occurs, you have to veer off the cutting line and saw to the edge of the workpiece, or turn off the saw and back the blade out of the cut. In either case, you must choose a new starting point for the cut. The key to avoiding such pitfalls is to visualize the cut before you make it so you can select the best starting point. If a dead end seems unavoidable, mark cutting lines on both sides of the workpiece. Occasionally, starting a cut on one side of a workpiece and finishing it on the other is the only way to make a cut.

On many contour cuts, making a series of straight "release" cuts through waste areas as illustrated below will greatly facilitate the procedure. If backtracking out of a cut is unavoidable, try to start with shorter cuts and back out of these, rather than beginning with the longer cuts. For particularly tight curves, drill a hole at the tightest curves and then cut to the hole along the marked cutting line.

One of the peculiarities of the band saw is that its blade will readily follow a marked line when cutting across the grain, but will tend to veer off when following the grain.

For greater control and accuracy, start a curve with a cut that runs across the grain rather than with it. When entering a curve from a straight cut, remember to reduce feed speed slightly to help ensure precision.

Cutting a Curve Freehand

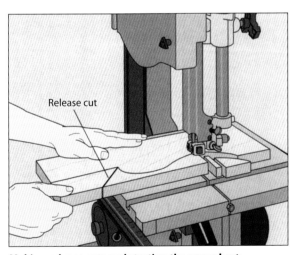

Making release cuts and starting the curved cut
To keep the blade from binding in the kerf of a curved cut, make a series of straight release cuts from the edge of the workpiece to the cutting line. The exact location of the cuts is arbitrary, but try to make them to the tightest parts of the curve, as shown. To start the curved cut, align the blade just to the waste side of the cutting line. Feed the workpiece steadily into the blade using your right hand, while guiding it with your left hand *(above)*. Make sure that neither hand is in line with the blade.

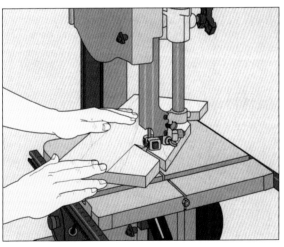

Finishing the cut
To cut the tightest parts of the curve, pivot the workpiece on the table, shifting your hand position as necessary. For the cut shown, saw to the end of the curved portion of the cutting line, feeding with your left hand. Pivot the workpiece with your right hand to avoid twisting the blade; veer off the cutting line and saw to a release cut, if necessary. Keep two fingers of your right hand braced against the table to maintain control of the cut *(above)*. Turn the workpiece around and cut along the straight portion of the cutting line.

Making Multiple Curved Cuts

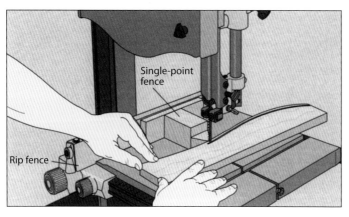

Setting up the fence and starting the cut
To produce multiple curved pieces with the same width from a single workpiece, cut the first curve freehand *(page 107)*. Then, make a T-shaped single-point fence with a rounded nose at the base of the T. Cut a notch in the base so that the guide assembly can be lowered to the workpiece. **(Note: In this illustration, the guide assembly is raised for clarity.)** Install the rip fence and screw the single-point fence to it with the tip of the base parallel to the blade. Position the rip fence for the width of cut. To start each cut, butt the workpiece against the tip of the single-point fence and feed it into the blade using both hands *(left)*. Keep the workpiece square to the tip of the single-point fence and ensure that neither hand is in line with the blade.

Finishing the cut
As the trailing end of the workpiece nears the tip of the single-point fence, shift your left hand to the back of the table to support the cut piece. Brace your left arm on the fence and hook two fingers over the edge of the table to keep your arm clear of the blade *(right)*. Continue feeding with your right hand until the cut is completed.

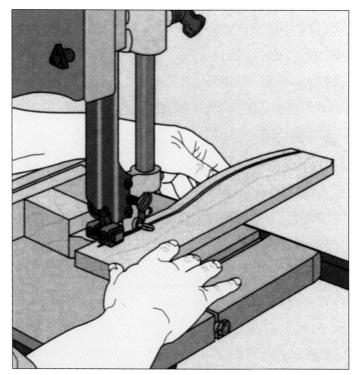

Pattern Sawing

Setting up a double-point fence

To cut the same curved pattern from different workpieces, cut the first piece freehand *(page 107)*; then, use it as a template to cut the other pieces. Prepare a double-point fence with a shallow notch at the end for the blade and a deeper notch below for the workpiece to slide under it. Screw the fence to an L-shaped support board that hugs the side of the table, then clamp the support board to the table, making sure the blade fits into the end notch of the fence. Use strips of double-sided tape as shown to fasten each workpiece to the template, ensuring that the straight edges of the boards are aligned. Trim the workpiece if necessary to prevent it from hitting the fence when you make the cut.

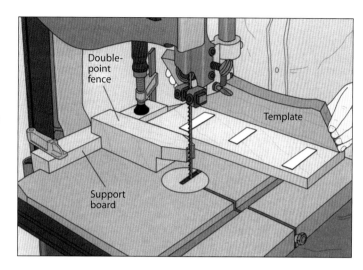

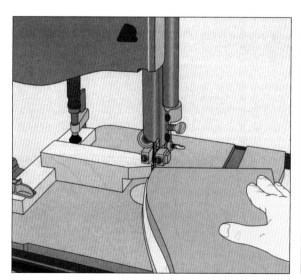

Lining up and starting the cut

Align the template and workpiece so that the edge of the template is parallel to the blade *(above)*. To begin the cut, use your left hand to feed the workpiece into the blade. Once the blade begins cutting, apply slight pressure with your right hand to press the template squarely against the end of the double-point fence. Keep the template in contact with both points of the fence throughout the cut.

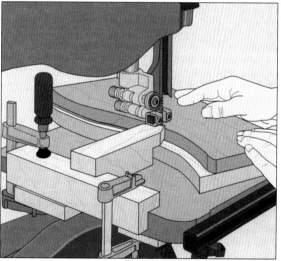

Completing the cut

Continue feeding with your left hand while using your right hand to keep the template flush against both points of the fence *(above)*; the template should ride along the fence as the blade cuts through the workpiece. Once you have finished the cut, pry the workpiece and template apart.

Rounding Corners

Setting up a quarter-circle-cutting jig
Cut a sheet of ¾ inch plywood slightly larger than the saw table, then feed it into the blade to cut a kerf from the middle of one side to the center. Clamp the sheet in position as an auxiliary table. Align a carpenter's square with the back of the blade gullets and mark a line on the auxiliary table that is perpendicular to the kerf. Then, mark a pivot point on the table the same distance from the blade as the radius of the rounded corners you plan to cut *(right)*. Cut another plywood sheet as a jig base and mark a square at one corner, with sides the same length as the radius of the rounded corners. Bore a hole for a screw at the marked corner (the spot marked "pivot point" on the inset illustration). Screw guides to adjacent edges of the jig base, then screw the jig base to the auxiliary table, centering the screw hole over the pivot point. Leave the screw loose enough to pivot the jig on the table. Round the marked corner of the jig by pivoting it into the blade *(inset)*.

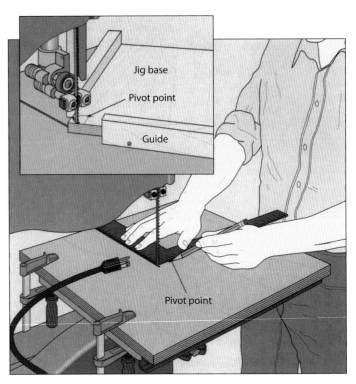

Rounding a corner
To round the corner of a workpiece, turn off the saw and seat the workpiece against the guides of the jig. Turn on the saw, then use your right hand to pivot the jig, feeding the workpiece into the blade; your left hand should hold the workpiece snugly against the guides. Round each corner of a workpiece the same way *(left)*.

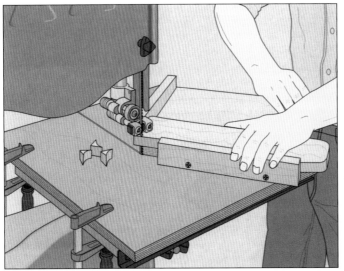

Circle-Cutting Jig

For cutting perfect circles, use a shop-built circle-cutting jig custom-made for your band saw. Refer to the illustration at right for suggested dimensions.

Rout a ⅜-inch-deep dovetail channel in the middle of the jig base, then use a table saw to rip a thin board with a bevel along two edges to produce a bar that slides smoothly in the channel. (Set the saw blade bevel angle by measuring the angle of the channel edges.) Cut out the notch on the band saw, then screw the support arms to the underside of the jig base, spacing them far enough apart to hug the sides of the band saw table when the jig is placed on it. Bore two screw holes through the bottom of the dovetail channel in the jig base 1 inch and 3 inches from the unnotched end; also bore two holes into the bar as shown.

To prepare a workpiece for circle-cutting, mark the circumference and center of the circle you plan to cut on its underside *(far right)*. Then, use the band saw to cut off the four corners of the workpiece to keep it from hitting the clamps that secure the jig to the table as the workpiece pivots. Make a release cut from the edge of the workpiece to the marked circumference, then veer off to the edge. Turn the workpiece over and mark the contact point where the blade touched the circumference.

Screw the narrow side of the bar to the center of the workpiece through one of the bar's holes. Do not tighten the screw; leave it loose enough to pivot the workpiece. Then, slide the bar into the channel and pivot the workpiece until the marked contact point is butted against the blade. Screw through one of the holes in the jig base to secure the pivot bar to the base.

To use the jig, pivot the workpiece into the blade *(left)*, feeding with your right hand and guiding with your left hand until the cut is completed.

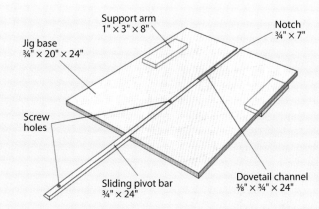

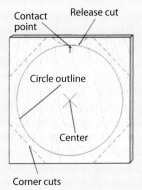

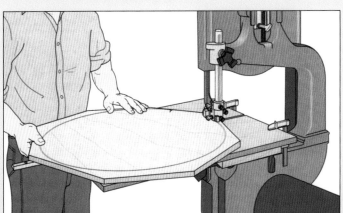

Straight Cuts

With a depth of cut that can be extended to 12 inches on some machines, the band saw is the ideal shop tool for resawing. Whereas a 10-inch table saw would take two passes to resaw a 6-inch-wide board, a standard 14-inch band saw can make the same cut in a single pass.

Because the band saw blade is relatively thin, it produces a narrower kerf—and less waste—than is possible with a table or radial arm saw. Resawing can be done freehand, but for more precision, use a pivot block and a featherboard *(page 114)*. The band saw is also an excellent choice for ripping narrow or round stock *(page 113)*.

Because the thin, flexible band saw blade has a natural tendency to pulse back and forth and sway from side to side imperceptibly as it cuts, you will need to keep your machine carefully tuned to get smooth and accurate cuts. Without such fastidious maintenance, crosscutting and ripping will be imprecise.

Band saw blades also have a tendency to "lead," or veer away from a straight line during a cut. This effect can be minimized by reducing feed speed and using sharp blades that are properly tensioned and tracked *(page 101)*.

Although more pronounced with narrower blades, some blade lead is generally unavoidable. However, the lead of a particular blade is usually constant and predictable, so you can usually angle your rip fence to compensate for it.

Crosscutting is a safe procedure on the band saw. But remember, one of the shortcomings of this machine is that crosscutting is limited by the width of the throat: typically 10 to 14 inches on a two-wheel consumer-grade tool.

> **Shop Tip**
>
> **Compensating for blade lead**
> To set the angle of the rip fence and ensure accurate, straight cuts when using the rip fence as a guide, adjust the position of the fence on the saw table for each blade in the shop. Mark a cutting line on a board that is parallel to its edge. Then, cut halfway along the line freehand. You may have to angle the board slightly to keep the blade on the line; this is the result of blade lead. Mark a line on the table along the edge of the board. Align the rip fence parallel with this line whenever using the same blade.

Ripping

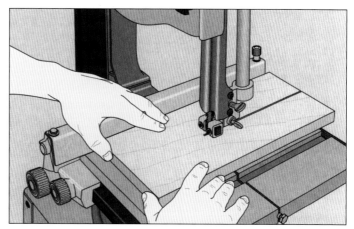

Ripping a board
Position the rip fence for the width of cut, adjusting its angle to compensate for blade lead. Butt the workpiece against the fence and feed it steadily into the blade with the thumbs of both hands *(above)*. To maintain proper control of the cut, straddle the fence with the fingers of your left hand and keep three fingers of your right hand braced on the table. Make sure that neither hand is in line with the blade.

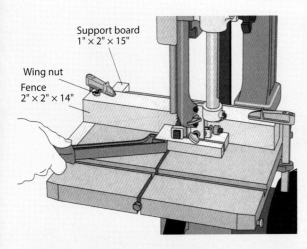

A Shop-Made Rip Fence

Like a commercial fence, the rip fence shown at left can be adjusted to compensate for blade lead. First, fasten a wooden fence to a support board with a bolt and wing nut. The board should rest flush against the front edge of the saw table. Ensure that the fence will pivot when the wing nut is loosened.

To use the fence, first mark a line on the table for the blade lead *(page 112)*. Hold the support board in position, then loosen the wing nut to pivot the fence and align its edge with the marked line. Tighten the wing nut, then clamp the fence in place. Feed short or narrow stock, as shown, using a push stick.

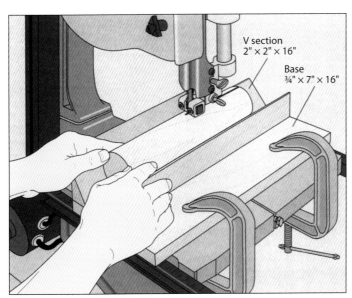

Ripping a cylinder

Rip a cylinder using a shop-made V-block jig. First, make the V section of the jig by bevel cutting *(page 116)* a 2-by-2 diagonally. Then, screw the two cut pieces side-by-side to a base of solid wood or ¾-inch plywood to form a V. To provide clearance for the blade when using the jig, make a cut halfway across the center of the V and the base.

To make the rip cut, slip the blade through the clearance cut, then clamp the jig to the table. Feed the cylinder into the blade using the thumbs of both hands *(left)*. Keep your fingers away from the blade. For a cylinder that is too narrow to be cut through from the front of the table without endangering your thumbs, stop feeding midway through the cut. Then, move to the back of the table to pull the cylinder past the blade.

Resawing

Using a pivot block and featherboard
To resaw a board, make a pivot block from two pieces of wood joined perpendicularly, with the shorter piece trimmed to form a rounded nose. Install the rip fence and screw the pivot block to it so that the rounded tip is aligned with the blade *(inset)*. Position the rip fence for the width of cut and adjust its angle to compensate for blade lead *(page 112)*. To start the cut, feed the workpiece into the blade using the thumbs of both hands; use your fingers to keep the workpiece flush against the tip of the pivot block. A few inches into the cut, stop feeding and turn off the saw. Clamp a featherboard to the table, propping it on a wood scrap to support the middle of the workpiece. Turn on the saw and continue the cut *(right)* until your fingers reach the featherboard.

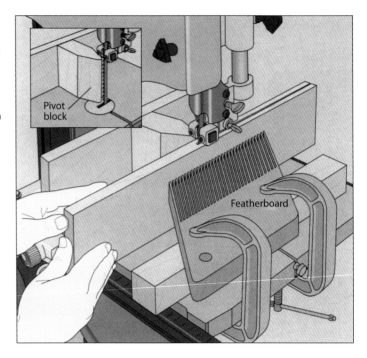

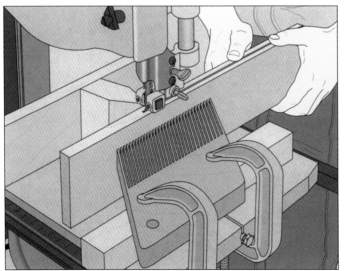

Completing the cut
With the saw still running, move to the back of the table to finish the cut. Use one hand to keep the workpiece square against the pivot block while pulling it past the blade with the other hand *(left)*.

Crosscutting

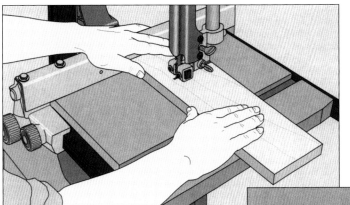

Using the rip fence as a guide
Position the rip fence for the length of cut, adjusting its angle to compensate for blade lead *(page 112)*. Butt the edge of the workpiece against the fence and feed it into the blade with the thumbs of both hands *(left)*. To maintain control of the cut, straddle the fence with the fingers of your left hand while keeping the fingers of your right hand braced on the face of the workpiece. Be sure that neither hand is in line with the blade.

Using the miter gauge as a guide
Use a carpenter's square to ensure that the miter gauge is perpendicular to the blade. Mark a cutting line on the leading edge of the workpiece. Holding the workpiece flush against the gauge, align the cutting line with the blade. With the thumb of your right hand hooked over the miter gauge, hold the workpiece firmly against the gauge and the saw table; use your left hand to push them together to feed the workpiece into the blade *(right)*. (Note: Do not try to compensate for blade lead when using the miter gauge for crosscutting.)

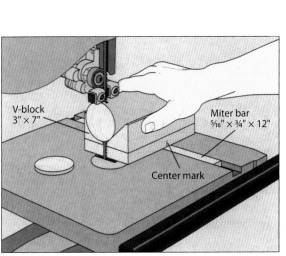

Crosscutting a cylinder
To crosscut a cylinder, make a V-block as described on page 113 but omit the clearance cut. Butt the V-block against the blade and mark the center of the miter slot on the base of the V-block. Screw a narrow strip of wood to the bottom of the V-block to serve as a miter bar, aligning the screws with the center mark; countersink the screws to keep them from scratching the saw table when using the V-block. Glue a sandpaper strip to the inside edges of the V-block to keep the workpiece from slipping during the cut. Insert the miter bar into the miter slot and seat the workpiece in the V-block so that it overhangs the edge of the V-block by an amount equal to the width of cut. Using your right hand to hold the workpiece firmly in the V-block, push it into the blade *(left)*.

Angle and Taper Cuts

By setting the band saw's miter gauge at an angle or tilting the saw table you can make precise angle cuts, such as miters, bevels and tapers. For a miter cut, use a sliding bevel to set the miter gauge to the desired angle—the gauge can be turned up to 90°—and then make the cut as you would a standard crosscut. For best results, make a test cut, check the angle of the cut edge with a square and adjust the miter gauge setting, if necessary.

For a bevel cut, tilt the table to the desired angle—band saw tables tilt up to 45° to the right and 10° to the left—and, for a cut along the grain, install the rip fence on the right-hand side of the blade. This will position the workpiece on the "downhill" side of the blade, keeping the workpiece—and your hands—from slipping toward the blade for a safer cut. For more accurate cuts, adjust the angle of the rip fence to compensate for blade lead *(page 112)*. Then, cut the bevel as you would a standard rip cut. The simple setups shown below can be useful for making multiple miter and bevel cuts.

Taper cuts can be made freehand, but for several identical pieces, using a jig *(page 117)* guarantees uniform results.

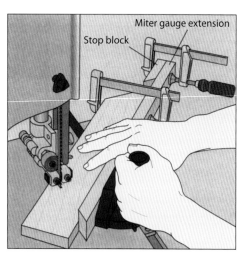

Making Repeat Angle Cuts

Mitering both ends of a board
Loosen the handle of the miter gauge and set the gauge to the desired angle. Then, screw a board to the gauge as an extension and cut off the end to the left of the saw blade. Glue a sandpaper strip to the extension to minimize the chance of the workpiece's slipping during a cut. Use the extension as a guide to cut the first miter, then make the miter cut on one end of a stop block. To cut the second miter, mark a cutting line on the leading edge of the workpiece. Holding the workpiece flush against the miter gauge, align the cutting line with the blade and butt the stop block against the end of the workpiece. Clamp the stop block and workpiece to the extension, then hook the thumb of your right hand over the miter gauge to hold the workpiece firmly against the gauge and the table. Use your left hand to feed the workpiece into the blade *(left)*.

Cutting bevels
Loosen the table lock knobs and set the saw table to the desired angle. Screw a board to the miter gauge as an extension and cut off the end of it. Use the extension as a guide to cut the first bevel. To cut the second bevel, mark a cutting line on the leading edge of the workpiece. Then, holding the workpiece flush against the miter gauge, align the cutting line with the blade and butt a stop block against the end of the workpiece. Clamp the stop block to the extension, then hook the thumb of your right hand over the miter gauge to hold the workpiece firmly against the gauge and the table. Use your left hand to push the miter gauge and workpiece together through the cut *(right)*.

Making Taper Cuts

Using a commercial taper jig
Install the rip fence to the right of the blade, then hold the taper jig flush against the fence. Pivot the hinged arm of the jig until the taper scale indicates the cutting angle—in degrees or inches per foot. Mark a cutting line on the leading edge of the workpiece, then seat it against the work stop and hinged arm. Position the fence so that the cutting line on the workpiece is aligned with the saw blade, then adjust the angle of the fence to compensate for blade lead. To make the cut, use the thumbs of both hands to slide the workpiece and the jig as a unit across the table, feeding the workpiece into the blade *(left)*. Use the fingers of your left hand to hold the workpiece against the jig, ensuring that neither hand is in line with the blade.

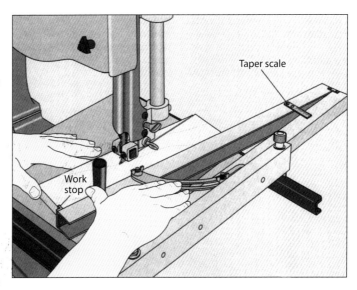

Taper Jig
Mark a line with the desired taper on the workpiece, then place the workpiece on a board with a perfectly square edge, aligning the marked line with the board's edge. Trace along the long edge of the workpiece to mark an angled cutting line on the board. Saw along the cutting line freehand, stopping 2 inches from the end of the cut at the bottom of the board. Turn the board 90° to cut out the lip. To use the board as a jig, set up the rip fence to the right of the blade, then hold the jig flush against the fence. Align the edge of the jig's lip with the saw blade and lock the fence in position, adjusting its angle to compensate for blade lead. Seat the workpiece against the jig. Use the thumbs of both hands to slide the workpiece and the jig as a unit across the table, feeding the workpiece into the blade. Use the fingers of your left hand to hold the workpiece against the jig, ensuring that neither hand is in line with the blade.

Cutting Duplicate Pieces

An effective method for producing multiple copies of the same shape is to fasten layers of stock together and cut the pieces in one operation with a technique known as stack sawing. Not only is it faster than cutting all the pieces separately; it also ensures that each piece is a precise copy of the original pattern. The method is possible because of the band saw's unique capacity to cut through very thick wood. With a 6-inch depth of cut a band saw can cut through as many as eight pieces of ¾-inch plywood in a single pass.

To bond the layers of wood together in preparation for the cut, some woodworkers drive nails through the waste area; others use clamps. Both methods, however, can be hazardous if the blade accidentally strikes a nail or a clamp. A safer way is to use double-sided tape to hold the pieces together temporarily.

A stop block on the saw table will also save time when you are crosscutting repeatedly to turn out duplicate pieces. With the setup shown below, you can speed the job of cutting a cylinder into identical slices.

Two Setups for Duplicate Pieces

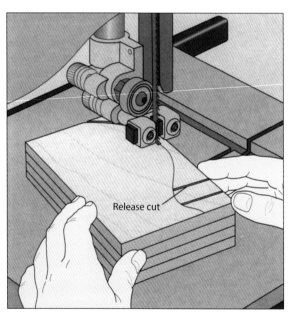
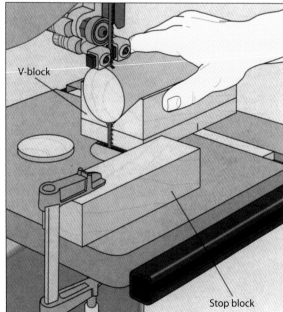

Stack sawing

Fasten the pieces together in a stack, then mark a cutting line on the top piece. Before turning on the saw, make sure that the blade is perfectly square with the saw table *(page 103)*; any error will be compounded from the top to the bottom of the stack. To cut the stack, first make any necessary release cuts *(page 107)*. For the curve shown, align the blade just to the waste side of the cutting line, then use the thumbs of both hands to feed the stack steadily along the marked path *(above)*. Keep your fingers on the edges of the stack and braced on the table to keep them safely away from the blade.

Using a stop block

Make a V-block with a miter bar as you would to crosscut a cylinder *(page 115)*. To produce several identical pieces, insert the V-block miter bar into the miter slot and clamp a stop block to the table so that the distance between the stop block and the blade equals the desired cut-off length. For each cut, seat the workpiece in the V-block and butt it against the stop block. Using your right to hold the workpiece firmly in the V-block, push them together to feed the workpiece into the blade *(above)*.

Woodworking MACHINES

Band Saw Joinery

A hallmark of fine craftsmanship, the dovetail joint is commonly used by cabinetmakers to join together corners of better-quality drawers and casework. The dovetail's interlocking pins and tails provide a joint that is not only strong and durable but visually pleasing as well.

Cutting dovetail joints on the band saw offers advantages over using either hand tools or other power tools. For all the artistry and uniqueness of handcrafted dovetails, the hand-tool approach is a laborious process. And while a router will make quick work of the job, it often produces pins and tails that are uniform in size and spacing. The result is a strong joint but one lacking in character.

Cutting dovetails on the band saw offers power-tool-type speed and precision. And as the following pages show, it is possible to tailor a dovetail joint on the band saw with the same flexibility you might bring to a handmade joint.

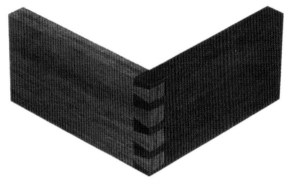

Dovetail joint

The sequence of operations is straightforward: First, outline the pattern of pins on one end of a pin board. Then, use a simple setup to cut all the pins on both ends of each pin board one after another. Once the waste is chiseled out, you can use the finished piece as a template for outlining the tails on the tail boards.

Making a Dovetail Joint

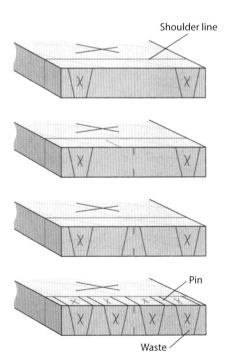

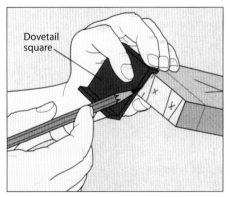

Marking the pins

Outline the pins for the joint, following the sequence shown in the diagram at left. First, mark the outside face of each workpiece with a big X. Then, set a cutting gauge to the thickness of the stock and scribe a line all around the ends of the workpieces to mark the shoulder lines of the pins. Next, use a dovetail square to outline the pins on an end of one workpiece, starting with half-pins at each edge; you want the narrow sides of the pins to be on the outside face of the workpiece. Outline the remaining pins (*above*), marking the waste sections with an X as you go along. There are no rigid guidelines for spacing the pins of a dovetail joint, but spacing them fairly evenly, as shown, makes for a strong and attractive joint.

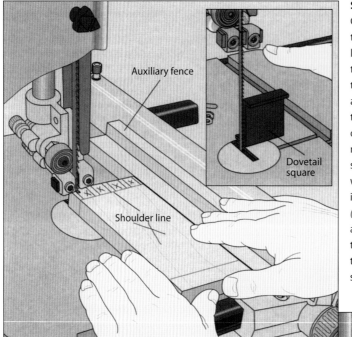

Setting up the table and making the first cut
Cut one edge of each pin with the saw table tilted downward to the right. To set up the table, loosen the lock knobs and set the table angle to match the edge of the dovetail square *(inset)*, then tighten the lock knobs. Set up the rip fence and screw a wooden L-shaped auxiliary fence to it. Then, with the workpiece on the saw table outside-face up, align the marked line for the right-hand edge of the first half-pin with the saw blade. Butt the auxiliary fence against the workpiece. To make the cut, feed the workpiece into the blade using the thumbs of both hands *(left)*; press the workpiece flush against the auxiliary fence with your left hand and straddle the fence with your right hand. Stop the cut and turn off the saw when the blade reaches the shoulder line on the face of the workpiece.

Using a stop block for repeat cuts
With the blade butted against the shoulder line, hold a stop block against the workpiece and screw it to the auxiliary fence *(right)*. To cut the right-hand edge of the first half pin at the other end of the workpiece, rotate the workpiece 180° and hold it flush against the auxiliary fence. Then, make the cut the same way you cut the first half pin, stopping when the workpiece touches the stop block. Rotate the workpiece 180° again, align the blade with the marked line for the righthand edge of the next pin, butt the auxiliary fence against the workpiece and cut to the stop block. Continue, shifting the position of the rip fence as necessary and cutting the right-hand edge of each pin on both ends of the workpiece.

Cutting the pins' left-hand edges
Cut the left-hand edge of each pin with the table tilted downward to the left. Use the dovetail square to set the table angle; remove the table stop, if necessary. Install the rip fence to the left of the blade and screw the auxiliary fence to it. Then, cut the left-hand edges of the pins the same way you cut the right-hand edges. Next, use a chisel to remove the waste between the pins. With the workpiece outside-face up on a work surface, strike the chisel with a wooden mallet to cut through the wood just to the waste side of the shoulder line. Then, hold the chisel square to the end of the workpiece to split off each waste section in thin layers. Remove about one-half of each section, then turn the workpiece over to remove the other half. Finally, pare the edges of the pins with the chisel.

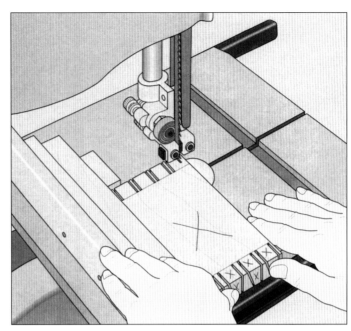

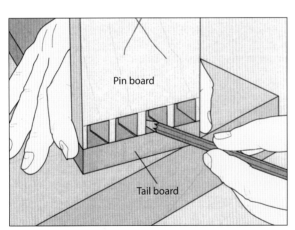

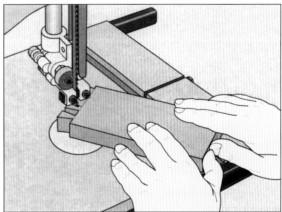

Outlining the tails
Set the tail board outside-face down on a work surface. Then, holding the pin board end-down on the tail board with its outside face away from the tail board, align the pins with the end of the tail board. Use a pencil to mark the outline of the tails on the ends of each tail board *(above)*, then mark the waste pieces.

Cutting the tails
Return the table to the horizontal position to cut out the waste between the tails. Cut the waste beside the half-tails at the edges of the workpiece with two intersecting cuts. For waste between tails, nibble at the waste with the blade, pivoting the workpiece as necessary to avoid cutting into the tails *(above)*. Test-fit the joint and make any necessary adjustments with a chisel.

Radial Arm Saw

Many woodworkers look on the radial arm saw as the table saw's poor cousin, suitable only for rough crosscuts and other carpentry-related chores. There are several reasons for this perception. One is that more woodworkers learn and fine-tune their craft on table saws than on radial arm saws. Table saws also have fewer moving parts and are easier to set up. Consequently, many woodworkers base their first impressions as a result of working on ill-adjusted machines.

There is no denying that the table saw is an excellent choice for the often repetitive chores of furniture making. Nevertheless, it is not the ideal stationary cutting tool for everyone. For repeat cuts, table saws usually require a jig to feed stock into the blade with uniform results. And some jigs are time-consuming to build or costly to buy. Table saws also need a fair amount of room to allow unhindered operation. Such space is at a premium in many home workshops.

With the exception of crosscutting very wide boards, radial arm saws can duplicate just about any job a table saw can perform. Even the crosscutting limitations are not severely restrictive. Most radial arm saws can rip up to a width of 25 inches, allowing you to cut a 4-foot-wide panel in half lengthwise. Moreover, the radial arm saw requires relatively little workshop space. Since stock either remains stationary on the saw table or passes laterally across the table, the machine can be stationed permanently against a wall.

A key advantage of the radial arm saw is that its blade remains visible as it cuts—a boon to safety. Another benefit is that most cuts can be made without having to shift the workpiece. Instead, the machine itself moves, with the arm pivoting on the column and the motor swiveling and rotating on its yoke. This allows the blade to be pulled through a workpiece at almost any angle. It also makes simple work of setting up the machine for custom work. But to take full advantage of the saw's capacity for cutting accurately, you must take the time to adjust the machine and keep it finely tuned.

On a poorly adjusted machine, the radial arm saw's flexibility can be its Achilles' heel. All of its sliding and pivoting movements must be highly controlled, and its movable parts must remain fixed when locked in position. Otherwise, the saw is condemned to a life of imprecise cuts. This holds true for any size machine. Radial arm saws typically range from 1 to 7 horsepower; the average home workshop model is rated at 1.5 horsepower. Blade size typically ranges from 8 to 24 inches; the standard home model has a 10-inch blade.

Augmented by a raised auxiliary table, the radial arm saw—its blade turned horizontally—cuts the notches for a finger joint.

This jig will allow you to make miter cuts on the radial arm saw with the blade in the standard cross-cutting position—90° to the table. The jig ensures that mating boards will form a perfectly square joint.

Woodworking **Machines**

Radial Arm Saw

Back to **Basics**

Anatomy of a Radial Arm Saw

The radial arm saw is essentially a circular saw suspended above a work table. For most operations, the blade cuts through the workpiece and runs along a kerf in a piece of hardboard or plywood that is glued to the saw table.

The machine's many pivoting and sliding parts enable it to carry the blade into a workpiece from a variety of different angles and directions. Sliding the yoke along the arm pulls the blade across the table for a crosscut. Swiveling the arm on the column allows for miter cuts; the maximum range of the model illustrated below is nearly 90° to the right and 50° to the left. Tilting the motor and blade makes a bevel cut possible, while rotating the yoke to bring the blade parallel to the fence sets up the machine for a rip cut.

Depending on the width of the stock you need to cut, two types of rip cuts are feasible: an in-rip, with the blade

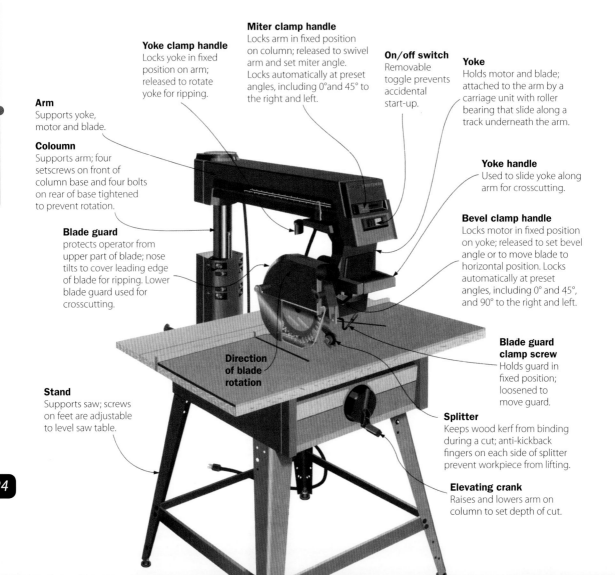

Yoke clamp handle
Locks yoke in fixed position on arm; released to rotate yoke for ripping.

Miter clamp handle
Locks arm in fixed position on column; released to swivel arm and set miter angle. Locks automatically at preset angles, including 0° and 45° to the right and left.

On/off switch
Removable toggle prevents accidental start-up.

Yoke
Holds motor and blade; attached to the arm by a carriage unit with roller bearing that slide along a track underneath the arm.

Arm
Supports yoke, motor and blade.

Coloumn
Supports arm; four setscrews on front of column base and four bolts on rear of base tightened to prevent rotation.

Yoke handle
Used to slide yoke along arm for crosscutting.

Bevel clamp handle
Locks motor in fixed position on yoke; released to set bevel angle or to move blade to horizontal position. Locks automatically at preset angles, including 0° and 45°, and 90° to the right and left.

Blade guard
protects operator from upper part of blade; nose tilts to cover leading edge of blade for ripping. Lower blade guard used for crosscutting.

Direction of blade rotation

Blade guard clamp screw
Holds guard in fixed position; loosened to move guard.

Stand
Supports saw; screws on feet are adjustable to level saw table.

Splitter
Keeps wood kerf from binding during a cut; anti-kickback fingers on each side of splitter prevent workpiece from lifting.

Elevating crank
Raises and lowers arm on column to set depth of cut.

Woodworking Machines

turned closest to the column, and an out-rip, with the blade swiveled farthest away from the column.

Although the blade is kept vertical to the table for most operations, it can also be tilted to operate horizontally. Such a position is particularly useful for tasks such as cutting grooves, finger joints and moldings.

Light and compact enough to move around the shop or travel to construction sites, this portable 8¼-inch radial arm saw can usurp the many roles of a table saw. Fitted with a special bit and equipped with an accessory motor shaft that turns at 18,500 rpm, this model will double as an overhead router.

Rip clamp handle
Locks yoke in position on arm for ripping and for some cuts with blade in horizontal position; released for crosscutting.

Dust spout
For dust collection system; adjustable nozzle directs dust away from work area.

Arm cover
Keeps dust from entering rear part of arm.

Motor
One end holds blade; opposite end serves as accessory shaft for attaching a variety of accessories.

Miter clamp adjustment screw
Turned to adjust tension on miter clamp; hole in arm cover provides access.

Fence
Prevents workpiece from moving during crosscutting; guides workpiece across table for ripping. Owner-installed. Usually set between front and rear tables as shown; positioned behind rear table when cutting wide stock.

Column adjustment bolts
Four bolts control amount of play between column and column base.

Rear table

Table clamp
Presses rear saw table and spacer flush against fence and front saw table.

Auxiliary table
Replaceable hardboard or plywood panel glued to front saw table; blade runs in kerfs cut in auxiliary table.

Column base cover

Table spacer
Removable to allow installation of a wider fence.

Radial Arm Saw — Back to Basics

Setting Up

The setup procedures described on these pages may seem long and involved, but do not neglect them. Without careful maintenance, your machine will not cut with precision. A problem with many radial arm saws is that adjustments are left too loose, allowing excessive play in moving parts and resulting in sloppy cuts.

Adjust the table the clamps *(below)* and the sliding mechanisms *(page 128)* before every new project. Each time you use the saw, clear the sawdust from the gap between the table and the fence, and clear the track underneath the arm. Periodically, touch up the moving parts with a silicone-based lubricant. It is also important to square the blade *(page 129)* and check for heeling *(page 130)*.

To test your adjustments, crosscut a 12-inch-wide board and a 1-by-3 standing on edge, then check the cut ends with a carpenter's square.

Adjusting the Table

Leveling the table with the arm
Tilt the motor until the arbor points down, its end slightly above table level. Then swivel the arm to position the arbor over the rail nuts on both sides of the table; in each position measure the gap between the arbor and the table. If the measurements are not equal, raise the low end of the table by turning the rail nut in a clockwise direction, using the head of an adjustable wrench to lever up the table surface *(right)*. Then make the same adjustment on the other side of the table.

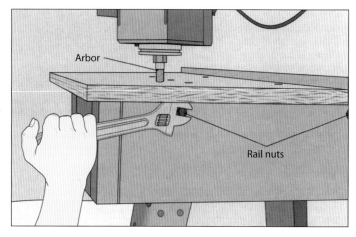

Adjusting the Clamps

Adjusting the miter clamp
Swivel the arm to the right to a position between 0° and 45°. Lock the clamp and try to push the end of the arm toward the 0° position *(left)*. If there is any play in the arm, adjust the clamp that holds it in place. For the model shown, you will need to use a hex wrench to tighten the miter clamp adjustment screw, located inside an access hole in the arm cover.

Fine-tuning the yoke clamp

Rotate the yoke to a position between the ones used for crosscutting and ripping. Lock the clamp, then use both hands to try to push the motor to the crosscutting position *(right)*. The motor should not budge; if it does, adjust the clamp that locks it in position. For the model shown, unscrew the knob from the yoke clamp handle and remove the wrench-like lower part of this device. Use the wrench to tighten the adjustment nut located under the arm by holding the upper part of the handle and pulling the wrench toward it *(inset)* until the two are aligned. Lock the clamp and check again for play. If necessary, tighten the nut further; otherwise, screw the knob back in place. This adjustment may vary on some models; check your owner's manual.

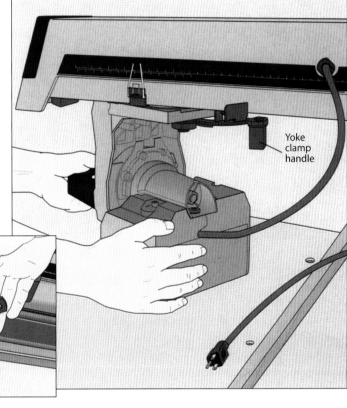

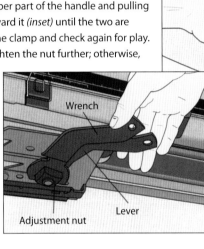

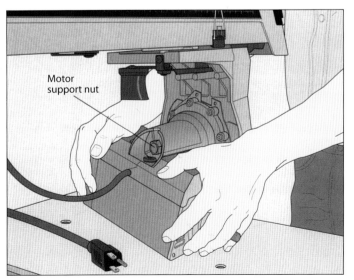

Adjusting the bevel clamp

Tilt the motor to a position between 0° and 45°. Lock the bevel clamp, then use both hands to try to move the motor *(left)*. If there is any looseness, adjust the clamp. For the model shown, use a socket wrench to tighten the motor support nut, then release the clamp and try tilting the motor to each of the preset angles; if you cannot move the motor, loosen the support nut slightly. Otherwise, lock the clamp again and check once more for play in the motor.

Checking the rip clamp

Lock the rip clamp, then use both hands to try to slide the yoke along the arm *(right)*. The yoke should not move; if it does, adjust the rip clamp. For the model shown, release the clamp, then use a wrench to tighten the nut at the end of the rip clamp bolt. Try sliding the yoke along the arm; if it binds, loosen the lock nut slightly. Otherwise, recheck the clamp and tighten the nut further if needed.

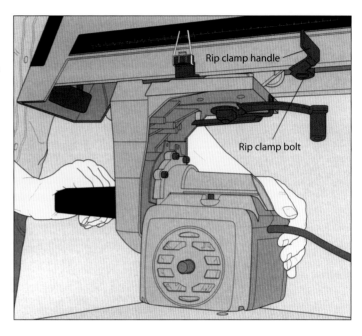

Caring for the Sliding Mechanisms

Adjusting the carriage roller bearings

Use a silicone-based lubricant to clean the track under the arm and the roller bearings to the front and rear of the carriage unit that attaches the yoke to the arm. To check the bearings, press your thumb against each one in turn while sliding the carriage away from your hand. The bearings should turn as the carriage slides along the arm. If your thumb keeps one of them from turning, you will need to tighten the bearing; if the carriage binds on the arm, a bearing will need to be loosened. In either case, loosen the bearing nut while holding the bolt stationary with a second wrench *(left)*. Tighten or loosen the bolt, as necessary, then retighten the nut. Adjust the other bolt by the same amount, then check the bearings once again.

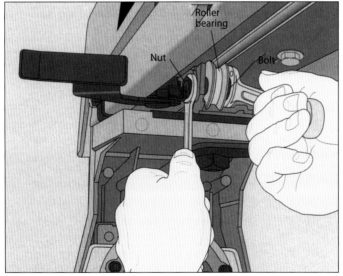

Woodworking Machines

Radial Arm Saw

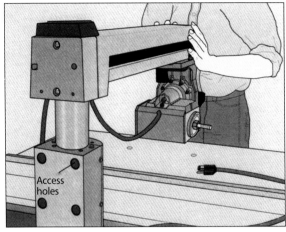
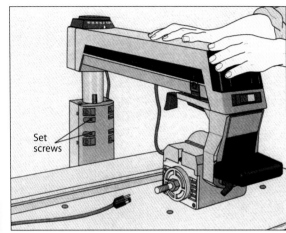

Adjusting column-to-base tension
Wipe the column clean, then loosen the four setscrews on the front of the column base using a hex wrench. To check column-to-base tension, use both hands to try to lift the end of the arm *(above, left)*; there should be little or no give to the column. Turn the elevating crank in both directions; the arm should slide smoothly up and down. If there is excessive movement at the column-to-base joint or if the arm jumps or vibrates as it rises and lowers, adjust the four bolts located in the access holes on the cover of the base. Repeat the tests and, if necessary, make additional adjustments. Then try pushing the arm sideways *(above, right)*; if there is any rotation of the column, tighten the setscrews just enough to prevent movement. Run through the tests a final time, fine-tuning the adjustments.

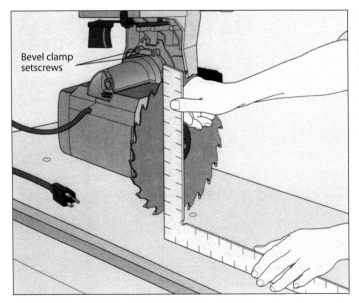

Squaring the Blade

Squaring the blade with the table
Set the yoke in the crosscutting position and install a blade *(page 133)*. Release the bevel clamp and tilt the motor counterclockwise as far as it will go in the 0° position. Then relock the clamp. To check the blade position, butt a carpenter's square between two teeth *(left)*. The square should fit flush against the side of the blade. If any gap shows between them, release the bevel clamp. Then, loosen the bevel clamp setscrews and tilt the motor to bring the blade flush against the square. Holding the motor in this position, have a helper lock the bevel clamp and tighten the setscrews. Tilt the motor to the 45° bevel position, then return it to the 0° position and check the blade once again.

Setting the arm perpendicular to the fence
Release the miter clamp and swivel the arm to the right as far as it will go in the 0° position, then relock the clamp. Release the rip clamp and butt the two sides of a carpenter's square against the fence and the blade tooth nearest to the table. Holding the blade steady, slide the yoke along the arm *(right)*; pull slowly to avoid dulling the tooth. The blade should make a constant rubbing sound as it moves along the edge of the square. If a gap opens up between the blade and the square, or if the blade binds against the square as it moves, loosen the setscrews on the column base. To close a gap between the blade and the square, tighten the top right screw; to eliminate binding, tighten the top left screw. Once the arm is square to the fence, tighten the lower screws, alternating from left to right.

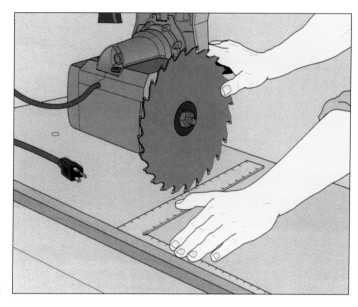

Correcting Blade Heel

Fine-tuning horizontal rotation
Install a blade *(page 133)* and set the motor in its horizontal position; tilt the motor counterclockwise as far as it will go, then lock the bevel clamp. To test for heeling—blade rotation that is not parallel to the table—build an L-shaped sounding jig and bore two holes in it. Sharpen the ends of two dowels and fit them into the jig as shown. Then position the jig to align a blade tooth near the back of the table directly over the vertical dowel. Lower the blade until the tooth rests lightly on the dowel; clamp the jig in place. Wearing a work glove, spin the blade backward and listen *(left)*. Next, slide the yoke along the arm to align a tooth near the front of the table over the dowel and repeat the test. The sound should be the same in both positions. If it is not, release the bevel clamp and loosen the two screws on either side of the motor support nut. Repeat the tests until the sound stays the same, then lock the bevel clamp and tighten the screws.

Woodworking Machines

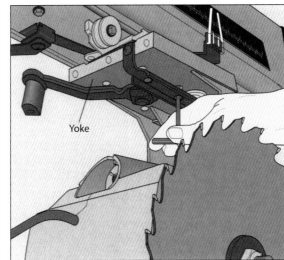

Eliminating vertical heeling

Tilt the motor counterclockwise as far as it will go in the vertical position, then lock the bevel clamp. To test for vertical heeling, position the sounding jig so that the tip of the horizontal dowel aligns with a blade tooth near the back of the table. Lower the blade and send it spinning backward so you can sample the sound *(above, left)*. Slide the yoke along the arm and repeat the process, once again listening for changes in tone. If there is a discrepancy, release the yoke clamp and loosen the four screws under the yoke using a hex wrench *(above, right)*. Rotate the motor as necessary and retest until each test produces a similar tone. Then, lock the yoke clamp and tighten the screws. To check your adjustments, crosscut a 12-inch-wide board and then a 1-by-3 standing on edge. Check the cut ends using a carpenter's square.

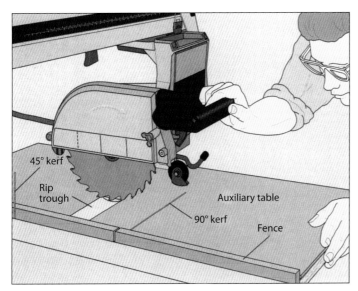

Installing a Fence and Auxiliary Table

Install a fence of ¾-inch-thick, knot-free wood between the table spacer and the front table. For an auxiliary table, cut a piece of ¼-inch hardboard or plywood the same size as the front table and use contact cement to glue it down, leaving a slight gap for sawdust. Before crosscutting or making miter cuts, slice through the fence and ¹⁄₁₆ to ⅛ inch deep into the auxiliary table in the 90° and 45° paths of the blade. Then, rotate the motor to the in-rip position *(page 140)* and pull the yoke along the arm to furrow out a shallow trough.

Radial Arm Saw Blades and Accessories

Like its shop cousin the table saw, the radial arm saw is only as good as the blade on its arbor. To get the best performance from your machine, keep its blades clean and in good repair. Inspect the arbor washers and blade collars, and replace any damaged parts. Use a rag to wipe sawdust or loose dirt from a blade; remove resin or pitch with steel wool and turpentine. Spray-on oven cleaner is also useful for dissolving stubborn deposits. To protect blades from damage, hang them individually on hooks or, if you stack them, place cardboard between them. Replace blades whenever they become cracked or chipped; sharpen non-carbide tipped blades regularly. A dull or damaged blade is more likely to contribute to accidents than a sharp blade in good condition.

In general, the radial arm saw uses the same types of blades as a table saw *(page 68)*. Combination blades are suitable for 90 percent of the jobs you will be doing. Blades for specific jobs, such as cross-cutting or ripping, are also available. In any case, it is important to consider the hook angle of a blade *(right)*. The larger the angle, the bigger the bite—and the greater the risk of a blade running across a workpiece when crosscutting or lifting stock when ripping. In both cases, feed the blade through the workpiece slowly and firmly. While a hook angle of 30° would be suitable for a table saw, the same blade on the radial arm saw could prove unsafe. The ideal hook angle for a radial arm saw is 15° or less.

Carbide-tipped blades are the choice of most woodworkers today. Although they cost more than the traditional high-speed steel blades and are more expensive to have sharpened, they hold their edge considerably longer and are capable of more precise cuts.

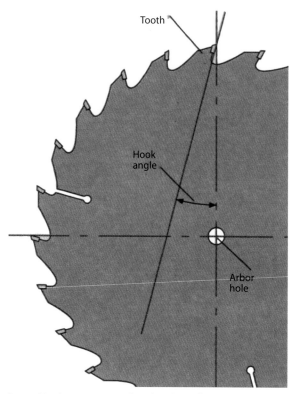

Formed by the intersection of one line drawn from the tip of a tooth to the center of the arbor hole and one drawn parallel to the tooth's face, blade hook angle determines how much bite a blade will have.

In addition to saw blades, the radial arm saw also accepts various accessories, which are attached to either the arbor or an accessory shaft at the opposite end of the motor. On some models, the shaft can spin at more than 20,000 rpm, making it ideal for powering router bits.

Woodworking Machines

Changing a Saw Blade

Removing and installing blades
Unplug the saw, lock the clamps and remove the blade guard. Then, fit one of the wrenches supplied with the saw on the arbor between the blade and the motor. Holding the arbor steady with this tool, use the other wrench to loosen the arbor nut. (Radial arm saw arbors usually have reverse threads; the nut is loosened in a clockwise direction.) Remove the nut and the outer blade collar, then slide the blade from the arbor. To install a blade, place it on the arbor with its teeth pointing in the direction of blade rotation. Install the collar and start the nut by hand. With one wrench on the arbor propped against the table, finish tightening the nut *(right)*, but avoid overtightening. Install the blade guard.

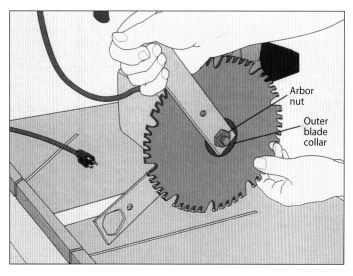

Setting the Blade Height

Preparing to make a cut
For a standard vertical cut, lower the blade into one of the precut kerfs in the auxiliary table *(page 131)*. For a cut partway through a workpiece, such as a dado cut, mark a line on the workpiece for the depth of cut, then set the stock on the auxiliary table and lower the blade to the line. For most saws, one turn of the elevating crank raises or lowers the blade ⅛ or ¹⁄₁₆ inch. To gauge the crank on your saw, lower the blade to within ½ inch of the auxiliary table, then crank in the opposite direction until the blade begins to move up. Hold a piece of scrap wood that is at least 1 inch thick against the fence and cut into it at one end. Turn off the saw and raise the blade by exactly one turn of the elevating crank. Slide the workpiece about ⅛ inch to one side and make another cut *(left)*. The difference in depth between the two cuts will show the amount the blade raises or lowers with one turn of the crank.

Safety

Although the many moving parts of a radial arm saw make it one of the most flexible machines in the workshop, they also make it one of the most dangerous. Crosscutting—the most basic use of the saw—requires you to pull the blade toward your body. And depending on the setting of the arm, yoke and motor, the blade can make its approach from several directions and angles. With every cut, you have to anticipate exactly where the blade will end up.

When you rip boards on a radial arm saw, you feed the workpiece into the blade, and this demands even greater care and concentration. The chances of kickback are high enough that the safety devices illustrated below and at right are absolutely essential.

Armed with a thorough knowledge of the machine's operation, you can approach it with a healthy mixture of caution and confidence—as you would with any other power tool in your workshop. Make certain that all the clamps for holding the arm, yoke, carriage and motor in position are locked whenever you turn on the saw. Also be sure to familiarize yourself with the owner's manual for your machine, and take the time to set up the many safety accessories and blade guards that are available. Remember, however, that no accessory or guard can compensate for a lack of careful attention and common sense.

For any cut, keep your fingers at least 6 inches away from the blade; use push sticks or featherboards where possible to feed or hold the workpiece. Wear safety glasses at all times, and a mask or respirator and hearing protection for extended use of the saw.

Ripping Safely

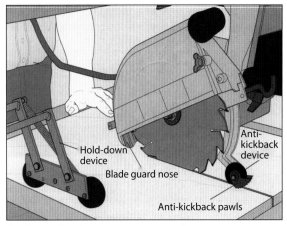

Setting up for the cut

Unplug the saw, then set the workpiece on the table so that you will be feeding against the direction of the blade's rotation. (Most blade guards have an arrow indicating which way the blade spins.) Follow the owner's manual instructions for setting the height of the anti-kickback device and for positioning the nose of the blade guard so that it just clears the workpiece. For added protection against kickback, install a wheeled hold-down device or spring-type hold-down fingers *(page 135)*. Feed the workpiece steadily *(above)*, making sure that neither hand is in line with the blade.

Radial Arm Saw Safety Tips

- Never install blades or other devices on both the arbor and the accessory shaft at the same time. Keep a safety screw cap or guard over the accessory shaft when it is not in use to prevent it from snagging hair or clothing.
- Never operate the saw without a blade guard. Use specialty guards for crosscuts and for molding or dado cuts with the motor and blade in the horizontal position. When making a rip cut, adjust the height of the anti-kickback device for the workpiece.
- Before starting a cut make sure that the motor is at full operating speed.
- Do not rip a workpiece that is shorter than 12 inches. When making a crosscut on stock shorter than 7 inches, use a hold-down device, rather than a hand, to secure the workpiece to the table or the fence.
- When ripping, ensure that the edge of the workpiece in contact with the fence is smooth and straight; feed from the side of the table opposite the splitter and anti-kickback device.
- To avoid kickback, always hold the workpiece securely against the table and fence when crosscutting.
- After making a crosscut, lock the rip clamp as soon as the blade is back behind the fence.

Anti-Kickback Devices and Specialty Guard

The hold-down device shown at left features rubber wheels that ride along the top of the workpiece, pressing it down against the table. The mechanism is installed at a slight angle so that the wheels also push the workpiece against the fence. The wheels and collar can be adjusted to accommodate workpieces of virtually any thickness. To help prevent kickback, the wheels are designed to rotate in one direction only. When the yoke is rotated to the out-rip position *(page 141)* and the workpiece is fed from the other side of the table, the wheels are swung around to turn in the opposite direction.

Another safety accessory for use in ripping and molding operations is the set of metal hold-down fingers shown at right. Clamped to L-shaped rods that extend over the fence on either side of the blade, the fingers push the workpiece down on the table. The rods can be adjusted to accommodate various sizes of stock. With the motor tilted to its horizontal position, a special guard covers the portion of the blade facing the front of the table. To use the guard, you must first make a cutout in the fence to allow the device's shield to be lowered onto the workpiece. Before turning on the saw, spin the blade by hand to ensure that the guard does not obstruct it.

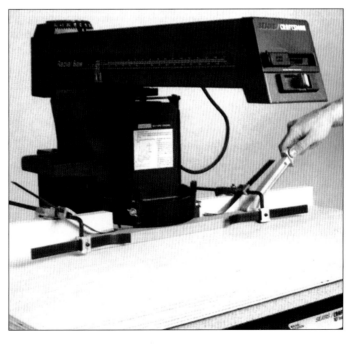

Crosscutting

The radial arm saw is best known for its convenience in crosscutting. The technique is straightforward: Hold the workpiece firmly against the fence and pull the yoke and the blade through the stock. Since the thrust of the blade is downward and toward the back of the table, the cutting action helps to keep the workpiece pressed against the table and the fence. However, several factors can cause the blade to climb up on the workpiece and jump toward you. These include a dull blade or one with teeth too large for the job at hand, poor quality wood, or loose roller bearings. But even with equipment in proper repair, it is still essential to remain in control of the blade at all times.

As a rule of thumb, hold the workpiece against the fence with your left hand, keeping it at least 6 inches from the blade; use a clamp to secure short stock *(page 138)*. With your right hand, pull the yoke, gripping it firmly to control the rate of cut. The slower the feed, the smoother will be the results. To cut several workpieces to the same length or to saw a thick workpiece in more than one pass, clamp a stop block to the fence, as shown below.

Making a Crosscut

Butt the workpiece against the fence with the 90° kerf in the fence lined up with the waste side of the cutting mark. Support long stock with roller stands or a table. Holding the workpiece snugly against the fence, turn on the saw, release the rip clamp and pull the yoke steadily through the cut *(right)* without forcing the blade. Once the blade cuts through the workpiece, push the yoke back, returning it to its place behind the fence. Lock the rip clamp.

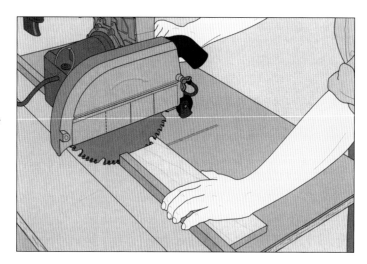

Making Repeat Cuts

Cut a small notch from one corner of the stop block, as shown, to prevent sawdust from accumulating between it and the workpiece. Measure along the fence to the left of the kerf the length of the piece you need to cut; clamp the block at that point. Butt the workpiece against the block and the fence, then make the cut *(left)*. To cut a thick workpiece in two passes, clamp the stop block to the fence and cut halfway through the stock, then flip the workpiece over and finish the cut.

Stop block

Angle Cuts

Miter, bevel and compound angle cuts can be made with the radial arm saw by tilting or angling its blade. The machine's arm swivels to the right or the left for miter cuts; the motor tilts clockwise and counterclockwise for bevel cuts. Compound cuts involve both swiveling the arm and tilting the motor. As discussed on page 139, you can also make a miter cut with a jig that holds the workpiece at an angle.

Both the arm and the motor have preset stops at 45° angles. To eliminate any play in these index settings, push the arm or motor as far as it will go in the stop positions and hold it there while you lock the clamp. To set the arm and motor at other angles, use a sliding bevel or the saw's miter and bevel scales for precise results. Always make a test cut first in a piece of scrap wood and measure the cut end with a protractor; then make any final adjustments.

Whenever possible, make miter cuts with the arm swiveled to the right, rather than to the left. Working on

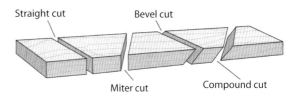

the left side of the table, you run the risk of pulling the blade beyond the table's edge. You often can make the same cut on the right side by turning the workpiece over.

Regardless of the type of angle cut, you first need to cut a kerf in the fence and the auxiliary table to provide a path for the blade. Make the kerf up to ⅛ inch deep for miter cuts, or deep enough for the blade teeth to be below the table surface for bevel or compound cuts.

Making Angle Cuts

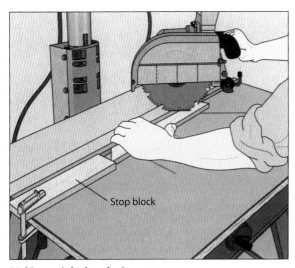

Making a right-hand miter cut

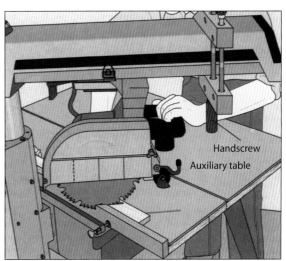

Making a left-hand miter cut

Making a bevel cut
Tilt the motor to the angle you need, raising the arm high enough to keep the blade from striking the table as it turns. Butt the workpiece against the fence with the waste side of the cutting mark aligned with the bevel angle kerf in the fence; if there is no such kerf on your machine, you will need to make one. Then, holding the workpiece snugly against the fence, pull the yoke steadily through the cut *(right)*. To make a bevel cut along the length of a workpiece, tilt the motor to the desired angle, then rotate the yoke to the in-rip position and make the cut *(page 140)*.

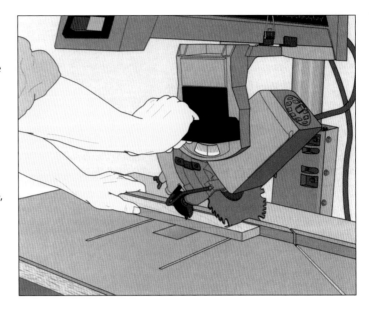

Shop Tip

Cutting a short workpiece
To cut a workpiece that is too short to hold safely by hand, secure it to the table with a toggle clamp. Screw the clamp to an auxiliary fence, then install the fence between the front table and the table spacer, making certain that the clamp will not be in the way of the blade. When you tighten the clamp, protect the workpiece with a wood block. To avoid lifting the fence out of its slot, do not overtighten.

Woodworking Machines

Miter Jig

To make 45° miter cuts without having to swivel the arm on the saw, use the shop-built jig shown at right. The jig holds the workpiece at an angle, so that the blade can remain in the 90° crosscutting position. Refer to the illustration for suggested dimensions.

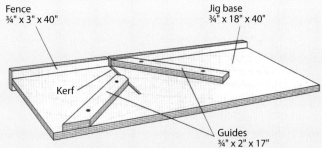

Fence
¾" x 3" x 40"

Jig base
¾" x 18" x 40"

Kerf

Guides
¾" x 2" x 17"

Before building the jig, make 45° miter cuts in the ends of two pieces of ¾-plywood that will serve as guides. Then cut the base and the fence and screw the two boards together, leaving enough of the fence protruding below the base to fit into the slot between the front auxiliary table and the spacer. Remove the standard fence and set the base on the table, sandwiching the jig's fence between the front surface and table spacer. With the blade in the 90° crosscutting position, slice through the jig fence and ⅛ inch deep into the base, pulling the yoke forward as far as it will go. Turn off the saw.

Screw one of the guides to the base so that its mitered end is flush against the fence with its point touching the kerf in the base. Position the mitered end of the second guide flush with the front of the table as shown. Use a carpenter's square to set the second guide at a 90° angle to the first one. Then, screw the second piece to the base, leaving enough space between the two guides for the stock you will be cutting to fit between them. Turn on the saw and pull the yoke across the kerf to trim off the corner of the second guide.

To use the jig, hold the workpiece flush against the right-hand guide, butting the end of the stock against the fence, and pull the yoke through the cut *(left)*. Next, hold the edge of the mating piece flush against the left-hand guide, with its end butted against the other guide. Pull the yoke through the cut. The resulting 45° ends should form a perfectly square joint.

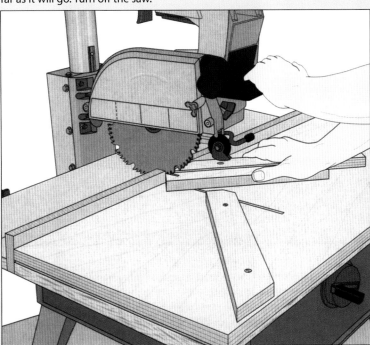

Ripping

Whether you are cutting with the grain of a piece of hardwood or softwood, or sawing along the length of a plywood panel with no defined grain pattern, ripping on a radial arm saw bears little resemblance to crosscutting. Rather than pulling the blade across a stationary piece of stock, you will be locking the yoke in a position that holds the cutting edge parallel to the fence and feeding the workpiece into the cut.

Depending on the width of the stock to be cut, the yoke can be rotated in two directions. For a narrow cut, typically up to 14 inches, the yoke is rotated to position the blade close to the fence. This is called the in-rip position. For wider stock, the yoke is rotated in the opposite direction, leaving the blade farther from the fence in the out-rip position. For the maximum width of cut, relocate the fence behind the rear table and use the out-rip configuration.

Because of the risk of kickback and the fact that you will be feeding the workpiece with your hands, ripping demands great care. Try to stand to one side of the stock as you feed it to the blade and keep your hands at least 6 inches from the cutting edge. Use a push stick to feed narrow stock or to complete a cut on a wide workpiece. Where possible, use featherboards to hold stock firmly against the fence.

As illustrated below, a good hold-down device will provide an additional measure of safety. Plan your cuts safely *(page 134)*, always feeding the workpiece against the direction of blade rotation: from the right-hand side of the table for an in-rip and from the left-hand side for an out-rip. Use the blade guard when ripping; and it is a good idea to install a new fence to keep the workpiece from catching in old kerfs.

Ripping a Board (In-Rip)

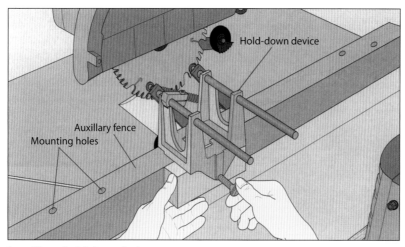

Installing a hold-down device

Unplug the saw, release the yoke clamp and rotate the yoke to the in-rip position; lock the yoke clamp. To position the blade for the width of cut you need, release the rip clamp, slide the yoke to the appropriate distance from the fence, and relock the clamp.

To install a wheeled hold-down device, replace the standard fence with an auxiliary fence about 1 inch thick and slightly higher than the thickness of the workpiece. Use the template supplied with the device to bore three sets of holes along the top edge of the fence; one set should be directly in line with the blade and the others to either side of the first. Fit the pins on the bottom of the hold-down device into one set of holes and tighten the thumbscrew using a wood block to distribute the pressure evenly along the fence.

Making the cut

Set the workpiece up against the right-hand side of the table. Standing to one side of the stock, slip its leading edge under the wheels of the hold-down device and feed it into the blade, applying pressure between the fence and the cutting edge. Make sure that neither of your hands is in line with the blade. When your fingers come within 6 inches of the blade, continue feeding with a push stick *(right)*. If you are using a hold-down device, move to the outfeed side of the table and pull the workpiece through; otherwise, finish the cut from the infeed side of the table. Retract the push stick carefully to prevent it from getting caught.

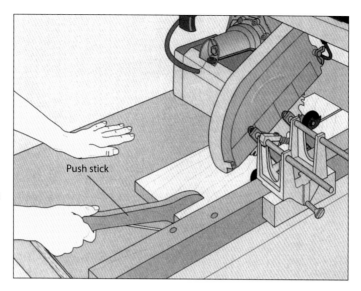

Using the Out-Rip Configuration

Cutting a panel to width

Unplug the saw, release the clamp and rotate the yoke to the out-rip position, with the blade away from the fence, as shown. Lock the yoke clamp. Position the blade for the width of cut. Move the fence behind the rear table, if necessary. Install a hold-down device, following the manufacturer's instructions to reverse the wheel-locking mechanism. Set up roller stands or a table to support the workpiece as it comes off the table.

To make the cut, lay the panel on the left-hand side of the table to allow you to feed against the direction of blade rotation. Butting the edge of the stock against the fence, slowly feed it into the blade. Apply enough lateral pressure to keep the panel flush against the fence *(left)*.

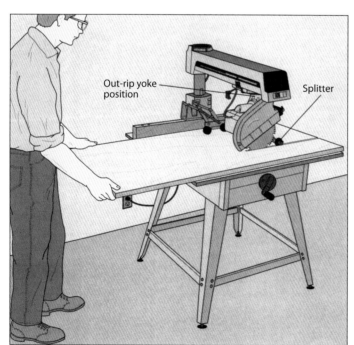

Dado Cuts

The same range of dado cuts that can be made on the table saw *(page 84)*—the cross-grain dado, the groove, the stopped groove and the rabbet—are also possible on a radial arm saw. As you will discover in the pages that follow, the ability of the radial arm saw to function in either vertical or horizontal planes means that there is often more than one way to make the same cut. Generally, most woodworkers find it easiest to keep the blade in the 90° crosscutting position when making cross-grain dadoes, rabbets along the ends of stock and grooves in wide boards. Moving the blade to the horizontal position works best for a rabbet along the edge of a workpiece or for a groove in a narrow board.

There is a way of cutting grooves without a dado head. With a standard saw blade you can make cuts on both edges of the groove and then saw out the waste between them in as many passes as is necessary. But the job can be done more quickly and precisely with a dado head mounted on the arbor. The radial arm saw accepts either the adjustable wobble dado head or the stacking dado. Although the stacking dado head is generally more expensive and takes longer to install, it produces cuts with flatter bottoms and smoother edges.

The wider swath cut by the dado chippers and blades, compared to the relatively narrow width of a standard saw blade, means that you will have to feed the stock more slowly. For safety's sake, keep track of the dado head during a cut, noting its location on the table when the workpiece hides it from view. Install a standard guard or a dado head blade guard when the blades are turned horizontally.

Installing a Dado Head

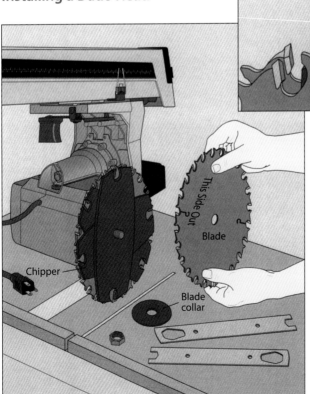

Adding blades and chippers
Remove the blade from the arbor *(page 133)* and install a dado head following the manufacturer's instructions. For the stacking dado shown, fit a blade on the arbor with the teeth pointing in the direction of blade rotation. Then install a chipper with its teeth centered in the gullets between two blade teeth. Fit on additional chippers, off-setting their teeth from those already in place. Put the second blade on the arbor *(left)*, making sure that its teeth do not touch those of the chipper resting against it *(inset)*. Install the blade collar and nut, keeping the blades and chippers carefully arranged as you do so. If you cannot tighten the nut all the way down, remove the collar. Install a standard guard or a dado head guard.

Cutting Dadoes with the Blades Positioned Vertically

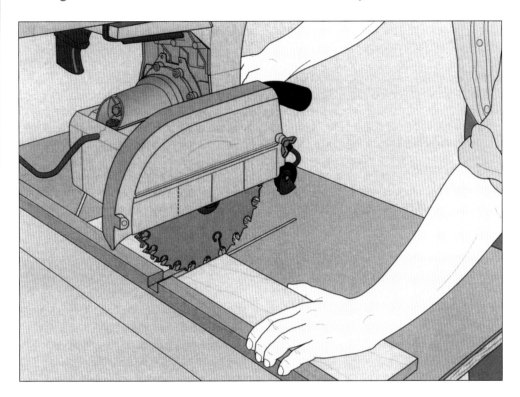

Making the cut
With the dado head in the 90° cross-cutting position, cut a kerf through the fence as deep as the dado you are planning to make. Mark two sets of cutting lines on the workpiece: one on its face to show the width of the dado, and one on its leading edge to show the depth. Butt the marks on the edge of the stock against the dado head and lower the blades and chippers to the appropriate depth. Slide the yoke behind the fence. Align the cutting lines on the face of the workpiece with the kerf in the fence. Then, holding the workpiece snugly against the fence, pull the yoke steadily through the cut *(above)*.

Shop Tip

Cutting repeat dadoes
To cut a series of equally spaced dadoes, use the simple setup shown below. Make a kerf in the fence and cut the first dado, then slide the workpiece along the fence, measuring to position the second dado the desired distance from the first. Before making the cut, drive a screw into the fence, with the head of the screw butted against the left edge of the first dado. Then cut the second dado and slide the workpiece along until the left edge of the second dado butts against the screw head. Continue in this manner until all the dadoes are cut.

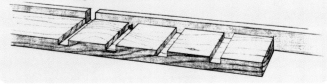

Cutting a Groove with Horizontal Blades and Chippers

Setting up the cut
With the dado head in the horizontal position, slide the yoke to the back of the table as far as it will go; lock all the clamps on the saw. Install an auxiliary fence or table *(page 145)*, then mark cutting lines on the workpiece to show the width and depth of the cut. Holding the workpiece against the fence, slide the yoke to align the dado head with the depth mark on the face of the stock *(right)*. To help keep track of the dado head's location when it is hidden by the workpiece, mark two lines on the table to delineate the cutting swath. Install a dado head guard, lowering its shield onto the workpiece. Spin the dado head by hand to make sure that it rotates freely. Slide the yoke behind the fence.

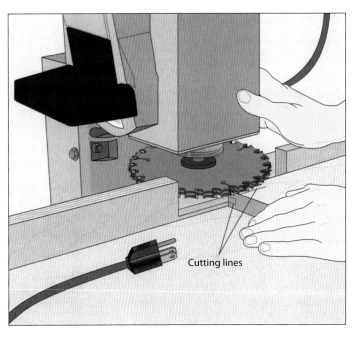

Cutting lines

Cutting the groove
Butt the workpiece against the fence, then clamp a featherboard to the table to hold the stock in alignment; clamp a support board at a 90° angle to the featherboard for extra pressure. Then, slide the workpiece back until you can lower the dado head to align its blades and chippers with the width marks on the end of the stock. Turn on the saw and use the thumbs of both hands to feed the workpiece steadily into the blades *(left)*; straddle the fence with the fingers of your right hand to help maintain control. To keep your hands from getting too close to the dado head, use a push stick to complete the cut.

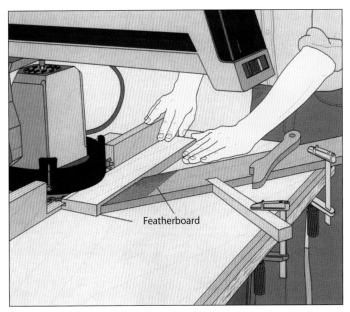

Featherboard

Woodworking Machines

Making a Stopped Groove

Setting up and starting the cut
Set up the cut as on the preceding page, but add one more set of cutting lines on the face of the workpiece to show the beginning and end of the groove. Standing on the right-hand side of the table, pivot the leading end of the workpiece away from the fence. Turn on the saw and align the cutting line for the beginning of the groove with the blade mark on the table surface *(right)*. Keeping both hands well clear of the dado head, hold the trailing end of the workpiece against the fence while pivoting the other end into the blades and chippers until the whole edge is flush with the fence.

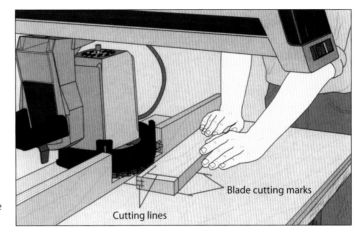

Cutting the groove
With your right hand gripping the trailing end of the workpiece, push the stock steadily forward. Use your left hand to keep the workpiece flush against the fence. Making sure that both hands stay well clear of the dado head, continue feeding until the cutting line for the end of the groove is aligned with the blade mark nearest you.

Finishing the cut
Slide your left hand carefully along the workpiece toward its leading edge, pressing the workpiece against the fence. Keeping both hands clear of the dado head, use your right hand to pivot the trailing end of the stock away from the fence *(right)*.

Drill Press

Originally designed for the metalworking trades, the drill press has found a second home in woodworking shops, where it has been a thoroughly welcome addition. Imagine trying to bore precise holes without it and you have an idea of how essential the tool is in exacting pursuits such as cabinetmaking. The drill press also does duty as a sander and mortiser and yet—despite its versatility—it takes up only a few square feet of workshop space and is relatively inexpensive. Many experts consider this machine a wise acquisition for the woodworker with limited space and budget.

One feature that distinguishes the drill press from other woodworking machines is its speed variability. Whereas power tools such as table saws are preset at the factory to operate at a single speed, the drill press can be adjusted for the job at hand. The range for a typical ½-horsepower motor extends from 400 to 4500 spindle revolutions per minute (rpm). Having the ability to vary the speed allows you to bore with equal efficiency through softwood and hardwood, ranging in thickness from a fraction of an inch to 3 or 4 inches thick. And as you see later on page 159, even this outer limit of drilling depth can be circumvented by means of a single shop-made jig.

Many drill press models include a chart on the inside of the belt guard that lists suggested speeds for boring through different stock, depending on the diameter of the drill bit. As a rule of thumb, the thicker the stock or the larger the drill bit diameter, the slower the speed used to bore the hole.

Some machines feature a knob that provides infinitely variable speed adjustments. On other machines, speeds are adjusted by shifting a belt to different steps on two pulleys. The model shown in this chapter features three pulleys and two belts, providing a range of 12 speeds in all.

Drill presses are rated according to the distance from the center of the chuck to the column, a factor that determines the widest workpiece a machine is capable of handling. A 15-inch drill press, for example, can cut a hole through the center of a work-piece that is 15 inches in diameter. The distance from the chuck to the column is one half that diameter, or 7½ inches.

Most drill presses for the home workshop are in the 11-to 16-inch range and are powered by ¼- to ¾-horsepower motors. Larger machines—20-inch models, for example—are more suitable for production shops and professional woodworkers.

Although the drill press is used primarily to bore holes, it can also perform other woodworking tasks, such as sanding curved surfaces.

Equipped with the appropriate jigs and accessories, the drill press can bore a variety of holes with a precision unmatched by hand tools. Here, a shop-made jig allows a woodworker to drill a series of angled holes in a rail. The holes will house and conceal the screws that connect the rail to a tabletop.

Woodworking Machines

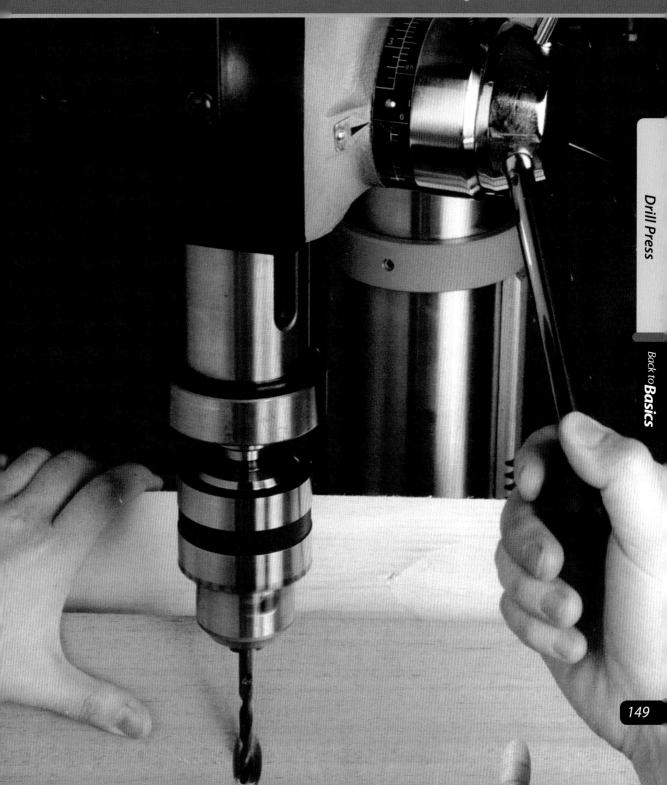

Drill Press

Back to Basics

149

Anatomy of a Drill Press

Drill presses come in various models and sizes, but the basic design is the same: A steel column 3 or so inches in diameter serves as a backbone to support a table and a motor that drives a spindle. The spindle is attached to a geared chuck whose jaws grip the shank of a drill bit or one of a variety of other accessories.

On some models, spindles are interchangeable. The standard spindle is mated to a chuck with a ½-inch capacity, which means that its jaws can accept shanks of drill bits and accessories up to ½ inch in diameter. Other spindles allow the drill press to accept router bits, molding cutters and mortising attachments.

The column is held upright by a heavy base, usually made of cast iron. For extra support and stability, the base can be bolted to the shop floor, but the weight of the drill press is normally adequate to keep it stationary.

The two most common types of drill presses are the floor model and the bench variety. The distinguishing feature is the length of the column: Floor models have columns from 66 to 72 inches high, whereas bench models range from 36 to 44 inches.

Since the table of a drill press can be positioned anywhere along the length of the column, floor models can handle longer workpieces. However, you can—to some extent—overcome the limitations of a bench-model drill press simply by swinging around the head of the machine. With the spindle extended beyond the edge of the workbench, the effective column length is the distance from the chuck to the shop floor.

While most drill presses have tables that tilt, the radial arm drill press features a head that rotates more than 90° right and left. Such tools can perform jobs impossible on conventional drill presses, including drilling through the center of a 32-inch-diameter circle.

Drill Press Belts and Pulleys

Spindle pulley
Turns spindle; features different steps to provide a range of speeds.

Jackshaft pulley
Intermediate pulley connected to spindle pulley so as to increase the range of speeds; driven by motor pulley.

Belt
Transfers power from motor pulley to jackshaft pulley; (other belt transfers power from jackshaft to spindle pulley).

Motor pulley
Driven by motor; connected by belt to drive jackshaft pulley. Features different steps to provide a range of speeds.

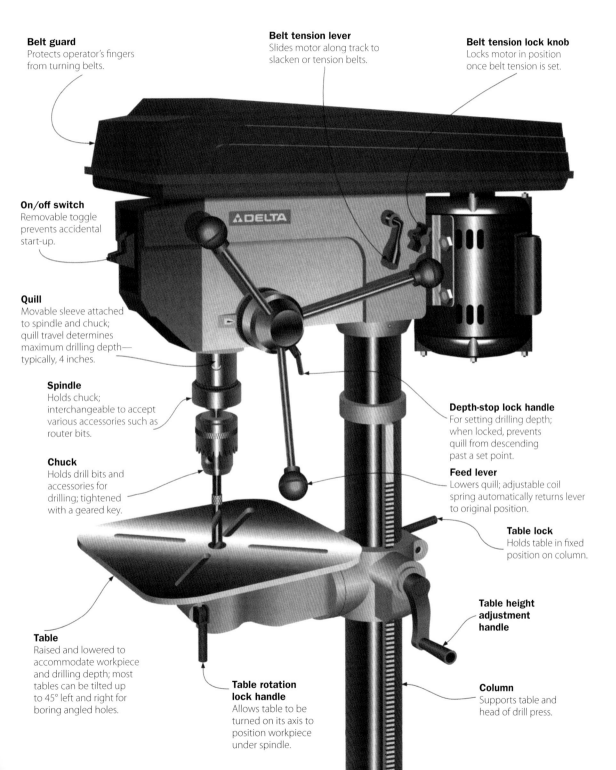

Set-up and Safety

Like any stationary power tool, the drill press has to be kept in adjustment to perform well. Before switching a machine on, check it carefully. Make sure all nuts and lock knobs are tightened. Even if you bought your machine new, there is no guarantee that it is perfectly ready to run. Check regularly that the table is square to the spindle.

There are also adjustments that have to be made depending on the particular job at hand, beginning with setting the drilling speed. The speed is changed either by turning a knob or by shifting the position of the belt—or belts—that connect the motor pulley to the spindle pulley.

The drill press has a reputation as a "safe" machine, and there is no denying that machines such as the table saw and jointer account for a greater number of serious accidents. Nevertheless, it is possible for even seasoned woodworkers to have accidents on the drill press. Unlike the table saw, a drill press will not kick back, but it can grip a small workpiece and send it spinning out of control if the stock is not clamped properly.

Setting the Drilling Speed

Changing belt position and setting belt tension
Loosen the belt tension lock knob and turn the belt tension lever counterclockwise to shift the motor toward the spindle pulley and slacken the belts. To set the desired rpm, position each belt on the correct steps of the pulleys, taking care not to pinch your fingers. (If your drill press has a drilling speed chart on the inside of the belt guard, refer to it in selecting the correct speed for the drill bit diameter you will be using and for the type and thickness of stock.) To set the belt tension, turn the tension lever clockwise while pressing the belt connected to the motor pulley until it flexes about 1 inch out of line *(right)*. Tighten the belt tension lock knob. Do not overtension the belt; this can reduce belt and pulley life. Undertensioned belts may slip.

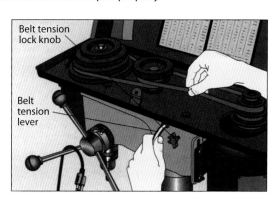

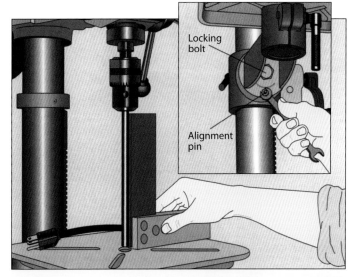

Squaring the Table

Aligning the table
Install an 8-inch-long steel rod in the chuck as you would a drill bit *(page 154)*, then raise the table until it almost touches the rod. Butt a try square against the rod as shown; the blade should rest flush against the rod *(left)*. If there is a gap, remove the alignment pin under the table using a wrench *(inset)*. Loosen the table locking bolt. Swivel the table to bring the rod flush against the square, then tighten the locking bolt. (Since the holes for the alignment pin will now be offset, do not reinstall the pin. The locking bolt is sufficient to hold the table securely in place.)

Correcting chuck runout

Use a dial indicator to see if there is any runout, or wobble, in the chuck. If there is, rap the rod with a ball-peen hammer *(right)* and then measure for runout again; 0.005 inch is considered the maximum acceptable amount. Pull the arm of the dial indicator out of the way each time you tap the rod.

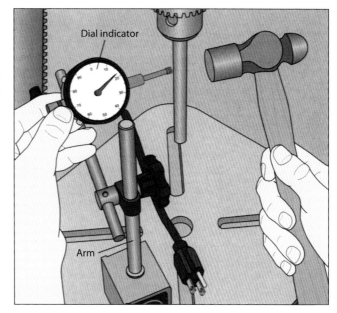

Replacing the Chuck

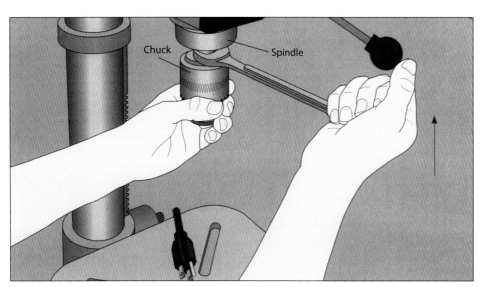

Removing and remounting a chuck

Chucks are commonly attached to the quill of a drill press with a tapered spindle. (Older models often have chucks that are simply screwed in place.) To remove a faulty chuck that features a tapered spindle, first lower the quill and lock it in place. Fit an open-end wrench around the spindle on top of the chuck and give the wrench a sharp upward blow *(above)*. The chuck should slide out. If not, rotate the spindle and try again. To remount the chuck, press-fit it into the spindle by hand. Then, with the chuck's jaws fully retracted, give the chuck a sharp blow with a wooden mallet.

Bits and Accessories

The range of accessories for the drill press is a testament to its versatility. In addition to a variety of sanding attachments, there are also bits for drilling $\frac{1}{32}$-inch holes, fly cutters for cutting 8-inch circles and plug cutters for making plugs and dowels.

Most drilling is done with twist or brad-point bits. Both consist of a cylindrical shank, which is held in the jaws of the chuck, and spiral-shaped grooves, known as flutes. The grooves allow waste chips and sawdust to escape from the hole, preventing overheating. The actual cutting is done by either sharp spurs or a cutting lip.

As with any cutting tool, drill bits must be sharp to work well. And like a saw blade, a drill bit is actually more dangerous when it is dull. A blunt bit has trouble digging into a workpiece and tends to heat up quickly, scorching the wood and the bit. Overheating can also result if drill bits are dirty or gummed up. Clean them with fine steel wool.

For any accessory you install in the drill press, be sure to remove the chuck key after tightening the jaws; otherwise, you risk launching a dangerous projectile once you turn on the machine. Some keys have a spring at the end of the geared segment. Pressure is required when inserting the key; once you let go, the key ejects automatically.

Changing a Drill Bit

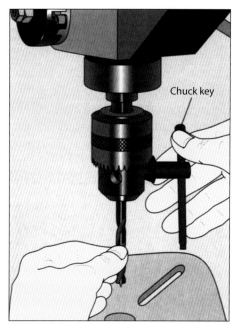

Removing and installing a bit
To remove a bit, use the chuck key to loosen the chuck jaws while holding the bit with your other hand. Slip the bit out of the chuck. To install a bit, open the jaws as wide as necessary, then insert the shank in the chuck. Steadying the bit to center it in the jaws, tighten the chuck by hand. Finish tightening using the chuck key *(above)*, fitting it in turn into each hole in the chuck. Remove the chuck key.

A Range of Bits and Accessories

Twist bit
The least expensive of commonly used drill bits; frequently sold in sets with a range of sizes.

Forstner bit
Bores perfectly flat-bottomed holes. Razor rim guides bit while chippers cut.

Brad-point bit
Produces cleaner holes than twist bit; does not "skate" off-line. Features a sharpened centerpoint and two cutting spurs.

Woodworking Machines

Drill Press

Column-Mounted Accessory Rack

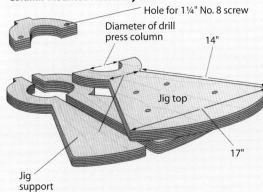

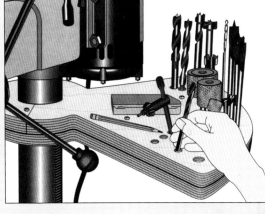

To save time searching for chuck keys and drill bits, use a shop-made storage rack. Cut two identical keyhole-shaped pieces of ¾-inch plywood to the dimensions shown above. Use a saber saw or coping saw to cut a circle out of each piece the same diameter as your drill press column. Then saw one piece in half lengthwise to serve as the jig support. The other piece will be the jig top; saw it across the circular cutout. Bore six screw holes for joining the top to its supports. Then, bore holes into the working surface of the jig to hold your bits and accessories shank-end down (above). Some woodworkers find it useful to have a small receptacle for odds and ends on the jig; a Forstner bit will make quick work of cutting such a hole. You will need a helper to hold the four pieces of the jig in place while you screw them together. Before doing that, however, make sure that the jig is turned so that it does not obstruct the rotation of the drill press's quill lever.

Plug cutter
For making small dowels and tapered plugs to conceal counterbored screws.

Hole saw
For boring large holes—typically, larger than 1½ inches. Available in models with fixed diameter or with adjustable blades. Pilot bit centers cutting edges.

Spade bit
For boring holes up to 1½ inches. Sharp centerpoint guides penetration, while flat blade slices into workpiece and removes waste.

Make sure you are familiar with your machine before attempting any work. Run through the drilling procedure before you turn on the machine, and never ignore the inner voice that warns you something may be amiss. Stop, check the setup again and continue the operation only when you are certain that what you are doing is safe.

Shop Tip

Checking table alignment
To check whether the table is square to the spindle, make a 90° bend at each end of a 12-inch length of wire coat hanger. Insert one end of the wire in the chuck and adjust the table height until the other end of the wire just touches the table. Rotate the wire; it should barely scrape the table at all points during the rotation, if not, remove the alignment pin under the table, loosen the table locking bolt and swivel the table to square it. Tighten the locking bolt.

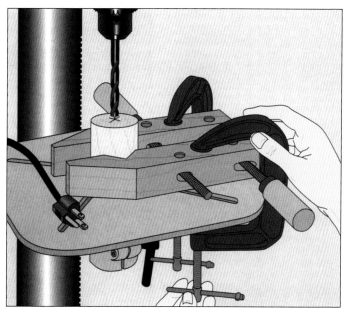

The Importance of Clamping

Choosing—or making—the right clamp
To prevent the drill bit from grabbing the workpiece and spinning it uncontrollably, always clamp small or irregularly shaped stock securely to the table before boring into it. When a conventional clamping setup does not work—as for the cylinder shown—improvise. Cut opposing V-shaped wedges out of a handscrew and clamp the cylinder in the handscrew, then use C clamps to secure the handscrew to the table *(left)*.

Straight and Angled Holes

Equipped with its tiltable table, the drill press can bore holes at virtually any angle. The steeper the angle, however, the more difficult it is for a brad-point or twist bit to dig into the stock without skating. Choose a Forstner or multispur bit when drilling holes at a very steep angle; both of these cutting accessories feature guiding rims that provide cleaner penetration.

Before drilling, make sure that the drill bit is lined up over the hole in the table: Otherwise, you risk damaging not only the bit but also the table itself. For further protection, some woodworkers also clamp a piece of wood to the drill press table.

For good results you will need to find the right combination of drilling speed *(page 152)* and feed pressure—the rate at which you lower the bit into the stock. Too much speed or feed pressure can cause burn marks on the workpiece and bit; too little will dull the bits cutting edge. With the proper combination, you should be able to cut steadily without having to put undue pressure on the quill feed lever.

Techniques for Basic Drilling

Setting up and drilling
To avoid splintering—particularly with plywood or particleboard—clamp a support board to the table and set the workpiece on top of it. Mark a starting point on the workpiece and align the bit over it. Rotate the feed lever steadily to feed the bit into the workpiece; use enough pressure to keep the bit cutting *(right)*. Retract the bit occasionally to clear the hole of wood chips, and if the machine labors or the wood starts to smoke, reduce the feed pressure or cut back on the drilling speed *(page 152)*.

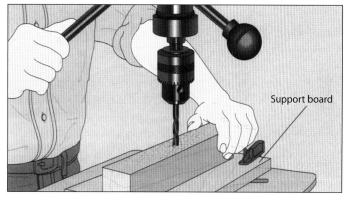

Boring Stopped Holes

Setting the drilling depth
For a stopped or blind hole—one that does not pass completely through a workpiece—mark a line at the desired depth of the hole on the edge of the stock. Then, lower the quill until the tip of the drill bit reaches the marked line. Hold the quill steady with one hand and, for the model shown, unscrew the depth-stop lock handle with the other hand and turn it counterclockwise as far as it will go *(right)*. Tighten the handle. This will keep the drill press from drilling any deeper than the depth mark.

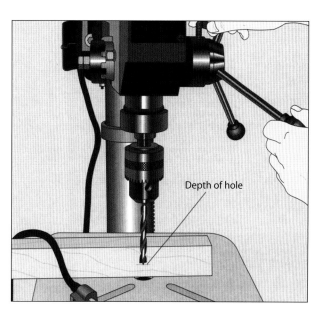

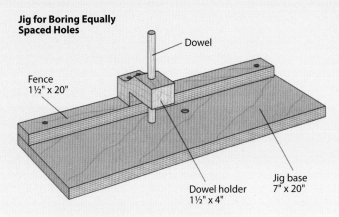

Jig for Boring Equally Spaced Holes
- Dowel
- Fence 1½" x 20"
- Dowel holder 1½" x 4"
- Jig base 7" x 20"

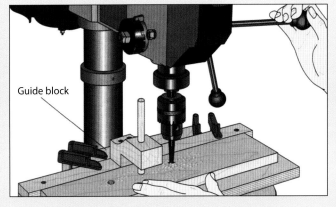

Guide block

To bore a row of uniformly spaced holes, make a shop-made jig to systematize the task, following the dimensions provided at left. Screw the fence to the jig base, flush with one edge, then attach a wood block at the center of the fence to serve as a dowel holder.

To use the jig, set it on the table of your drill press, then mark starting points on the workpiece for the first two holes in the series. Seat the workpiece against the fence of the jig and position the jig to align the bit—preferably a Forstner—over the first drilling mark. Butt a guide block against the back of the jig and clamp it to the table. If you are boring stopped holes, set the drilling depth *(left)*. Bore the first hole, then slide the jig along the guide block and bore a hole through the dowel holder. Fit a dowel through the hole in the holder and into the hole in the workpiece. Slide the jig along the guide block until the second mark on the workpiece is aligned under the bit. Clamp the jig to the table and bore the hole.

To bore each of the remaining holes, retract the dowel and slide the workpiece along the jig's fence until the dowel drops into the last hole you made then bore another hole.

Boring Angled Holes

Setting the table angle
Install a straight 8-inch-long steel rod in the chuck as you would a drill bit, then use a protractor to set the drilling angle you need on a sliding bevel. Loosen the table as you would to square it *(page 152)*. Then butt the bevel against the steel rod and swivel the table until the table rests flush against the handle of the bevel *(right)*. Remove the rod from the chuck and tighten the locking bolt. After installing the drill bit, set the drilling depth *(page 157)* to prevent the bit from reaching the table. For added protection, clamp a piece of wood to the table.

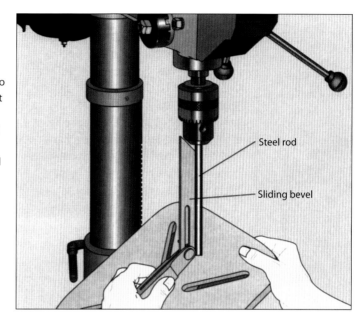

Tilting Table Jig

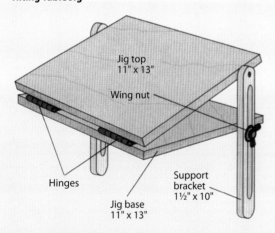

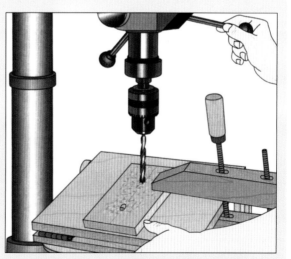

To bore angled holes without tilting the table, use a tilting jig, shopbuilt from ¾-inch plywood. Refer to the illustration above for suggested dimensions. Connect the jig top to the base using two sturdy butt hinges. Cut a ½-inch-wide slot in the support brackets, then screw each one to the top; secure the brackets to the base with wing nuts and hanger bolts.

To use the jig, center it under the spindle. Clamp the base to the table. Loosen the wing nuts and set the angle of the jig as you would the table *(above, left)*, but without removing the alignment pin or loosening the table locking bolt. Tighten the wing nuts, clamp the workpiece to the jig and bore the hole *(above, right)*.

Boring Deep Holes

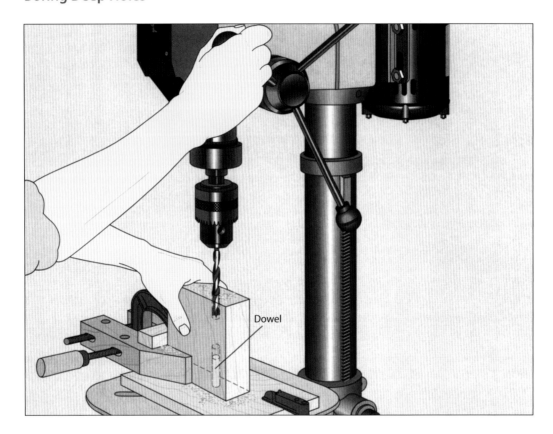

Exceeding the quill stroke
The maximum length that the quill can be extended—known as the quill stroke—limits most drill presses to boring no more than 4 inches deep at a time. To drill a deeper hole, use an extension bit or, if the hole is less than twice the quill stroke, perform the operation in two stages, as shown above. First, clamp a scrap board to the drill press table and bore a guide hole into it. Then, clamp the workpiece to the board and bore into it as deeply as the quill stroke will allow. Remove the workpiece and fit a dowel into the guide hole in the scrap board. Fit the hole in the workpiece over the dowel and bore into the workpiece from the other side. The dowel will ensure that the two holes in the workpiece are perfectly aligned.

Shop Tip

A simple center finder
Cut a 90° wedge out of a 7-by-12-inch piece of ¾-inch plywood. Screw a 12-inch-long 1-by-2 to the piece so that one long edge of the 1-by-2 bisects the wedge at 45°. To use the center finder, seat the workpiece in the wedge and use the 1-by-2 as a guide to draw a line across the diameter of the workpiece. Rotate the workpiece 90° and draw a second line across it. The two lines will intersect at the center of the workpiece.

Boring into Cylindrical Stock

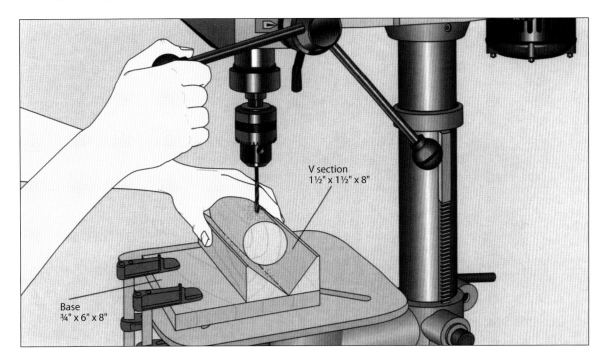

Using a V block
The safest way to bore into a cylinder is to secure it in a shop-made V-block jig. Make the V section of the jig by bevel cutting a 2-by-2 lengthwise using a table saw or band saw. Then, screw the two cut pieces to the base to form a V. Position the jig on the table so that the drill bit touches the center of the V when the quill is extended. Clamp the base to the table, seat the workpiece in the jig and bore the hole *(above)*.

Shop Tip

Drilling compound angles
To help line up entrance and exit holes when you are drilling at an angle, use this simple jig. Glue a 4-inch-long cylinder to a 5-by-10-inch piece of plywood. Clamp the base to the drill press table so that the cylinder is centered under the spindle and bore a hole into it. Sharpen one end of a 2-inch-long dowel, then fit the dowel into the cylinder. Mark both the entrance and exit holes on the workpiece and strike each mark with a punch. Position the exit punch mark on the dowel, hold the workpiece firmly and bore into the entrance punch mark. (**Caution:** Do not use this jig with stock too short to hold securely.)

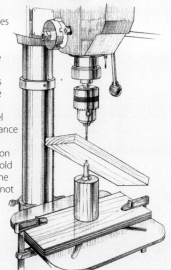

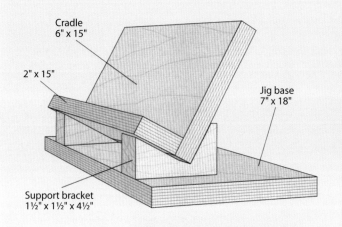

Pocket Hole Jig

Pocket holes are commonly used with screws to attach rails to a tabletop. They are drilled at an angle and solve the problem of having to screw straight through a 3- or 4-inch-wide rail. A pocket hole jig *(left, top)*, shop-built from ¾-inch plywood, makes simple work of such openings. For the jig, screw the two sides of the cradle together to form an L. Then cut a 90° wedge from each support bracket so that the wide side of the cradle will sit at an angle about 15° from the vertical. Screw the brackets to the jig base and glue the cradle to the brackets.

To use the jig, seat the workpiece in the cradle with the side that will be drilled facing out. Bore the holes in two steps with two different bits: Use a Forstner bit twice the width of the screw heads for the entrance holes and a brad-point bit slightly wider than the width of the screw shanks for the exit holes. (The wider brad-point bit allows for wood expansion and contraction.)

To begin the process, install the brad-point bit and, with the machine off, lower the bit with the feed lever, then butt the end of the workpiece against the bit. Position the jig to align the bit with the center of the bottom edge of the workpiece *(inset)*. Clamp the jig to the table and replace the brad-point bit with the Forstner bit.

Holding the workpiece firmly in the jig, feed the bit slowly to bore the holes just deep enough to recess the screw heads. Then, install the brad-point bit and bore through the workpiece to complete the pocket holes *(left, bottom)*.

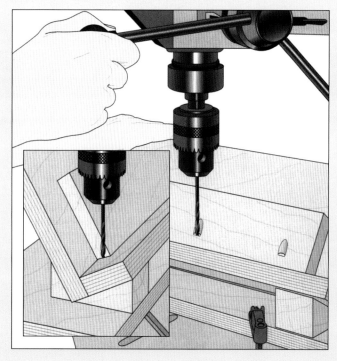

Woodworking Machines

Dowels, Plugs and Tenons

Dowels are 2- to 4-inch-long wood cylinders used to reinforce simple constructions such as butt joints in which two pieces of wood are butted together and held in place with glue. By drilling perfectly aligned holes in both pieces of such a joint and inserting dowels, you greatly strengthen the joinery. Another variation is the integral tenon, which looks and functions like a dowel but remains part of one of the wood pieces being joined.

The plug—a shorter cousin of the dowel—serves to conceal counterbored screws. Dowels and plugs can be cut from either softwood or hardwood. The difference between them—other than their length—is that dowels are cut from end grain to give them cross-sectional strength. Plugs, on the other hand, are not subject to any radial stress and can be cut either with or against the grain. They can either be concealed or used as a decoration depending on whether they are cut from the same stock as the workpiece.

Dowels of various diameters and in 3- or 4-foot lengths are widely available wherever wood is sold, but you can make your own if you outfit your drill press with a dowel

An integral tenon makes a strong joint and is relatively easy to cut. The tenon is produced with a dowel cutter at the end of a square piece of stock.

cutter. The best way to make plugs is to cut down a dowel or to use a plug cutter. With the latter accessory you can either cut through stock the same thickness as the plug or bore a stopped hole through thicker stock and pry the plugs out with a chisel.

Making Dowels and Integral Tenons

Using a dowel cutter
To cut dowels, clamp a block of wood to the table and bore into its end grain to the required depth with a dowel cutter *(far left)*. Free the dowels by cutting through the block with a table saw or a band saw. If you will be using the dowels for joinery, crimp their ends with the serrated jaws of pliers; this will provide the glue with an escape route and ensure proper glue coverage.

To cut an integral tenon on a long workpiece, tilt the table 90° and clamp the workpiece to the table, using pads to protect the wood. Also clamp a support board to the workpiece and to the table. Use a dowel cutter to bore to the required depth *(near left)*, then saw away the waste to expose the tenon.

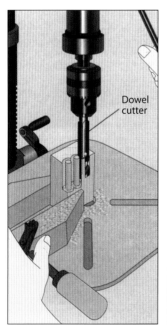
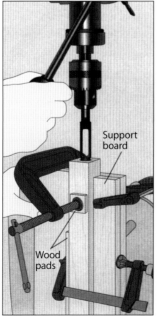

Mortising Techniques

Since the time of ancient Egypt, woodworkers have relied on the mortise-and-tenon joint to connect pieces of wood. Today, the joint is commonly used to join rails to legs on desks, tables and chairs. Like most joints, the mortise-and-tenon can be cut by hand. But for ease and efficiency in carving out mortises, the drill press equipped with a mortising attachment has become the tool of choice. The attachment consists of a bit that rotates inside a square-edged chisel. The bit cuts a round hole; the chisel then punches the corners square. The matching tenon can be cut easily on a table saw.

Chisels come in different sizes to cut mortises in a variety of widths. The depth is set with the drill press depth-stop; 7/8-inch is typical. As shown below, it is important to make sure that the attachment is adjusted to keep the workpiece square to the chisel. If you are cutting the mortise in round stock, use a V block to hold the workpiece securely in place.

The drilling speed *(page 152)* for mortising depends on both the type of stock and the size of the chisel. The larger the chisel, the slower the speed, especially when you are drilling into hardwood.

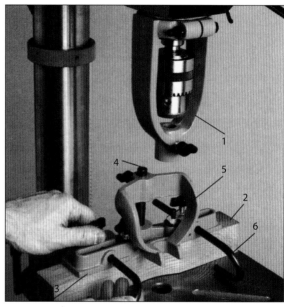

A typical mortising attachment consists of a chisel holder *(1)*, which is secured to the drill press quill by machine bolts at the top of the holder. The fence *(2)* and the hold-down bracket *(3)* on the table are held in place with screws, washers and wing nuts. The vertical bar *(4)* supports the hold-down arm *(5)*, which, along with the hold-down rods *(6)*, helps hold the workpiece firmly against the fence.

Installing the Chisel and Bit

Setting the gap between the chisel and bit
Insert the chisel into its holder and tighten the lockscrew. Push the bit up through the chisel into the chuck. Hold the tip of the bit level with the bottom of the chisel with a scrap of wood, then lower the bit by 1/32 inch. This will ensure proper clearance between the tip of the bit and the points of the chisel. Tighten the chuck jaws *(left)*.

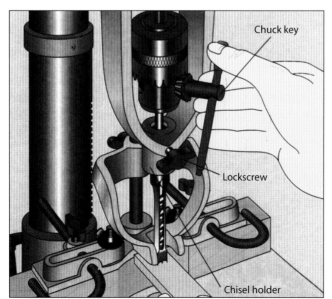

Squaring the Chisel

Adjusting the chisel
The chisel must be square to the mortising attachment fence or the mortises you cut will angle off-center, producing ill-fitting joints. To make sure that the chisel is properly aligned, butt a try square against the fence and chisel. The square should rest flush against both. If it does not, loosen the chisel holder lockscrew just enough to allow you to rotate the chisel and bring it flush against the square. Do not raise or lower the chisel while making the adjustment. Tighten the lockscrew *(right)*.

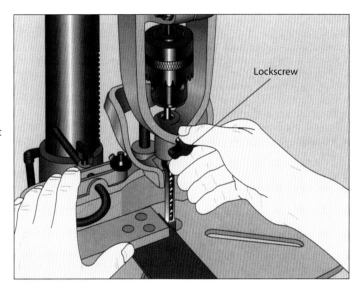

Cutting a Mortise

Setting up
Outline the mortise on the workpiece, centering the marks between the edges of the stock. To check whether the mortise chisel will be centered on the workpiece, butt a scrap board the same width and thickness as the workpiece against the mortising attachment fence and secure it with the hold-down rods. Bore a shallow cut into the board. Then, flip the board around and make a second cut next to the first. The cuts should be aligned. If not, shift the fence by one-half the amount that the cuts were misaligned and make two more cuts *(left)* to repeat the test. **(Note: Hold-down arm raised for clarity.)**

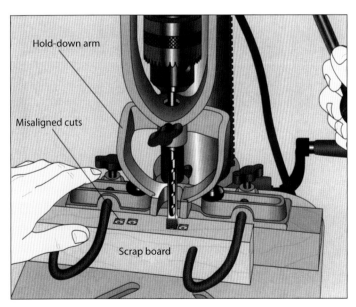

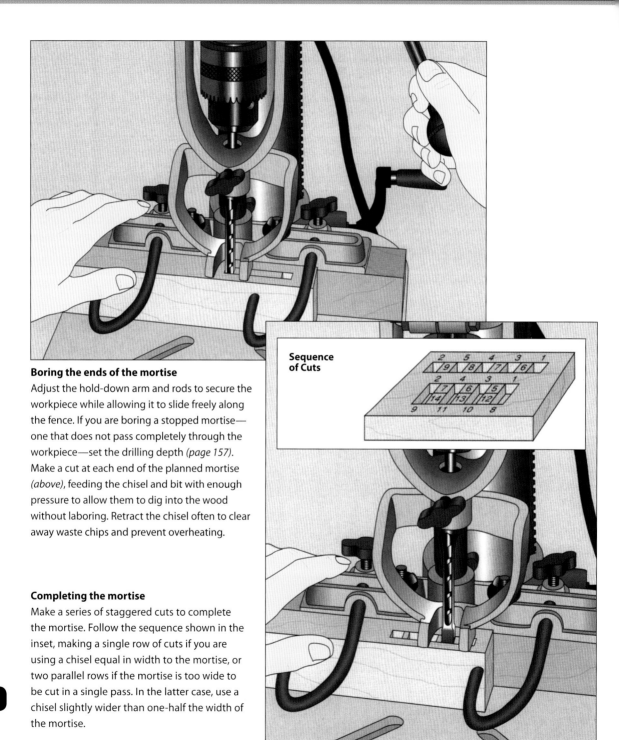

Boring the ends of the mortise
Adjust the hold-down arm and rods to secure the workpiece while allowing it to slide freely along the fence. If you are boring a stopped mortise—one that does not pass completely through the workpiece—set the drilling depth *(page 157)*. Make a cut at each end of the planned mortise *(above)*, feeding the chisel and bit with enough pressure to allow them to dig into the wood without laboring. Retract the chisel often to clear away waste chips and prevent overheating.

Completing the mortise
Make a series of staggered cuts to complete the mortise. Follow the sequence shown in the inset, making a single row of cuts if you are using a chisel equal in width to the mortise, or two parallel rows if the mortise is too wide to be cut in a single pass. In the latter case, use a chisel slightly wider than one-half the width of the mortise.

Woodworking Machines

The Drill Press as Sander

Drill presses make excellent sanders. The machine's table provides good support for the workpiece, holding it at 90° to the sanding drum to produce sanded edges that are square to adjacent surfaces. And with help from some simple jigs, the drill press can sand not only straight surfaces but curved ones as well.

Sanding drums come in diameters ranging from ½ to 3 inches. The shaft of a drum is inserted into the jaws of the chuck and secured in the same way that drill bits are installed. Sanding sleeves to cover the drum are available in a variety of grits—from a coarse 40 grit to a fine 220 grit. In most cases, sleeves are changed by loosening a nut at either the top or the bottom of the drum, which reduces the pressure and releases the sandpaper. Remove the old sleeve and slip on the new. Tightening the nut will cause the drum to expand and grip the sleeve securely.

As with standard drilling operations, sanding requires a variety of speeds depending on the job. The higher the rpm, the smoother the finish, but high speeds will also wear out your sleeves more quickly. Most sanding is done between 1,200 and 1,500 rpm. Sanding produces fine dust so remember to wear a dust mask.

In addition to sanding, the drill press can double as a router, although its relatively slow spindle speed keeps it from performing as well as its portable counterpart. While a drill press generates roughly 3,500 to 4,500 rpm, a router turns at more than 20,000 rpm, producing much smoother results.

To use your drill press as a router, you will need to buy a special spindle to attach router bits to the machine. Feeding the stock slowly will help compensate for the machine's slower speed.

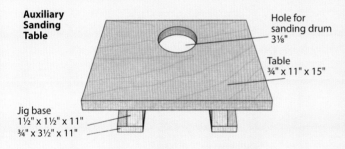

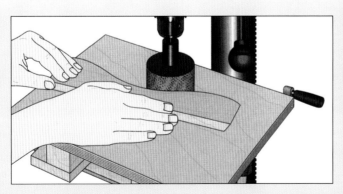

Auxiliary Sanding Table and Pattern Sanding Insert

Sanding drums larger than ⅞-inch in diameter are too wide to fit through the hole in most drill press tables. To make full use of the sanding surface of wider drums you will need to make a sanding table *(left, top)*.

Use a coping saw or saber saw to cut a hole in the plywood top, centering the opening 3 inches from the back of the table. Assemble the L-shaped jig base from 1-by-4 and 2-by-2 stock, then glue it to the table.

To use the jig, clamp the base to the drill press table with the circular hole directly underneath the drum. Adjust the table height to bring the bottom of the drum level with the jig.

Holding the workpiece firmly, feed it at a uniform speed in a direction opposite the rotation of the sanding drum *(left, bottom)*.

Jointer

The jointer may seem a rather pedestrian machine compared to the table saw or band saw, but any woodworker dedicated to precision and craftsmanship will attest that using this surfacing tool properly is the first step in turning rough boards into well-built pieces of furniture. The machine's main purpose is to shave small amounts of wood from the edges and faces of boards, yielding smooth, straight and even surfaces from which all subsequent measurements and cuts are made. The jointer gets its name from the fact that two edges run across its planing blades should fit together perfectly, forming a seamless joint.

Errors at the jointing stage of a project will have a ripple effect in all later procedures. Without a perfectly square edge to set against a table saw rip fence, for example, trimming a board to size will produce a flaw that will be further compounded when you try to cut a precise-fitting joint.

Traditionally, the task of creating smooth, square edges was performed with hand planes. Nowadays, woodworkers rely on the jointer to do the job more quickly, effortlessly and accurately. The machine functions much like an inverted hand plane with somewhat larger blades driven by a motor, addressing the workpiece from below rather than above.

The jointer is also useful in salvaging warped stock as well as in shaping rabbets, bevels and tapered legs *(pages 176–178)*.

Jointers are categorized according to the length of their cutterhead knives, which determines the maximum width of cut that the machine can make. Sizes for consumer models range from 4 to 8 inches; 6- and 8-inch jointers are the most popular. Depth of cut, which ranges from ⅛ to ½ inch, is another distinguishing feature. But unless you plan to make frequent use of the jointer's rabbeting capability, a shallow depth of cut is adequate: The typical bite seldom exceeds ⅛ inch.

When choosing a jointer, look for a machine on which the tables on both sides of the cutterhead are adjustable. And make sure the machine has a rigid, lockable fence that can be tilted for angle cuts.

The jointer is often confused with the planer *(page 179)*, but the two machines are not interchangeable. One important function of the planer that cannot be effectively performed by a jointer is planing a surface to make it parallel to the opposite surface. Planers can also handle wider stock, important when constructing panels such as tabletops.

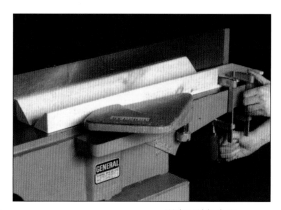

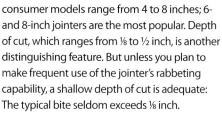

With a V-block jig clamped to the jointer infeed table, you can cut beveled edges into a workpiece accurately and safely.

In addition to smoothing stock or producing square edges, a jointer can be used to cut tapers in a workpiece, such as a table leg.

Woodworking **Machines**

Jointer

Back to ***Basics***

Anatomy of a Jointer

The jointer consists of infeed and outfeed tables separated by a cylindrical cutterhead. Cutterheads typically hold three knives and rotate at several thousand revolutions per minute. For a jointer to work properly, the outfeed table must be level with the knives at the highest point of their rotation. The model illustrated below has an outfeed table that is adjustable to keep it at the same height as the knives. For models on which the outfeed table is fixed, the knives must be raised or lowered to bring them to the proper height.

Depth of cut is determined by the amount that the infeed table is set below the outfeed table. The fence used to guide stock over the cutterhead is normally set at a 90° angle. But on most models the fence will tilt forward or backward for cutting bevels and chamfers.

Jointer Cutterhead

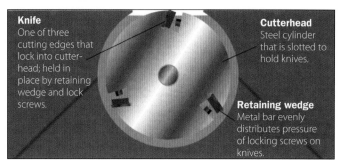

Knife
One of three cutting edges that lock into cutterhead; held in place by retaining wedge and lock screws.

Cutterhead
Steel cylinder that is slotted to hold knives.

Retaining wedge
Metal bar evenly distributes pressure of locking screws on knives.

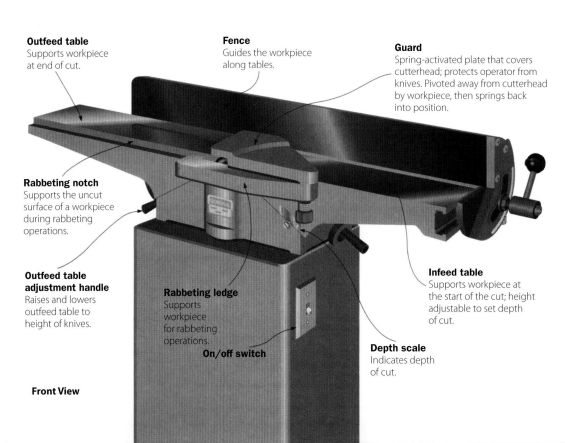

Outfeed table
Supports workpiece at end of cut.

Fence
Guides the workpiece along tables.

Guard
Spring-activated plate that covers cutterhead; protects operator from knives. Pivoted away from cutterhead by workpiece, then springs back into position.

Rabbeting notch
Supports the uncut surface of a workpiece during rabbeting operations.

Outfeed table adjustment handle
Raises and lowers outfeed table to height of knives.

Rabbeting ledge
Supports workpiece for rabbeting operations.

On/off switch

Infeed table
Supports workpiece at the start of the cut; height adjustable to set depth of cut.

Depth scale
Indicates depth of cut.

Front View

Woodworking **Machines**

Although the guard should always be left in place for standard operations, on most models it has to be removed for specialized work, such as rabbeting. On some machines, the guard can be installed behind the fence to provide protection during rabbeting work.

With a jointer on the right and a planer on the left, this machine combines two functions in a single appliance. The model shown can joint stock up to 6 inches wide and plane boards as wide as 12 inches.

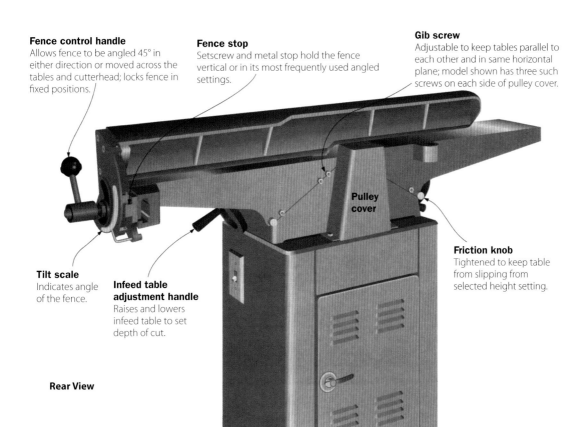

Fence control handle
Allows fence to be angled 45° in either direction or moved across the tables and cutterhead; locks fence in fixed positions.

Fence stop
Setscrew and metal stop hold the fence vertical or in its most frequently used angled settings.

Gib screw
Adjustable to keep tables parallel to each other and in same horizontal plane; model shown has three such screws on each side of pulley cover.

Pulley cover

Friction knob
Tightened to keep table from slipping from selected height setting.

Tilt scale
Indicates angle of the fence.

Infeed table adjustment handle
Raises and lowers infeed table to set depth of cut.

Rear View

Set-up and Safety

Accurate jointing depends on precise alignment of the two tables and the fence—the parts of the machine that guide a workpiece into and over the knives. Begin by ensuring that the outfeed table is at the same height as the cutting edges of the knives at their highest point. Then check that the tables are perfectly square to the fence and aligned properly with each other.

Before starting, make sure that the jointer is unplugged and install a clamp on the rabbeting ledge to hold the guard temporarily out of your way.

Once you have the machine properly tuned, pause and consider safety. The knives of a spinning cutterhead look seductively benign. It is easy to forget that this harmless-looking blur can cause as much damage to fingers and hands as can a table saw blade. Resist the temptation to operate the jointer without the guard in place. When the guard must be removed from its normal position in front of the fence for rabbeting operations, install it behind the fence if your jointer is set up for such a switch.

Even with the guard in place, always keep your hands away from the knives. When jointing the edge of a board, your hands should ride along the workpiece, rather than on the tables. When face-jointing, always use push blocks to feed a workpiece across the knives. Whatever the cut, remember to press the workpiece firmly against the tables and fence.

Setting Outfeed Table Height

Checking table height
Use a small wooden wedge to rotate the cutterhead until the edge of one of the knives is at its highest point. Then hold a straight hardwood board on the outfeed table so that it extends over the cutterhead without contacting the infeed table *(right)*. The knife should just brush against the board. Perform the test along the length of the knife, moving the board from the fence to the rabbeting ledge. Repeat the test for the other knives. If one knife fails the test, adjust its height as you would when installing a blade *(page 175)*. If none of the knives touches the board, adjust the height of the outfeed table.

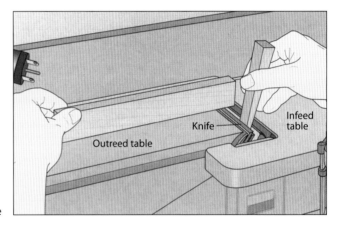

Adjusting the outfeed table height
Keeping the hardwood board over the cutterhead, turn the outfeed table adjustment handle *(left)*, raising or lowering the table until the edge of a knife just brushes against the board. Then check the table height in relation to the other knives.

Woodworking Machines

Aligning the Tables and Fence

Aligning the tables

Remove the fence, then use the adjustment handle for the infeed table to bring it to the same height as the outfeed table. Use a straightedge to confirm that the two tables are absolutely level. If the alignment is not perfect, adjust one or more of the gib screws at the back of the jointer until the straightedge rests flush on both tables; remove the pulley cover, if necessary, to access the screws. To adjust a screw, first loosen its locknut, then make the adjustment using with a hex wrench *(right)*. Tighten the locknut. At this point, the depth scale *(page 170)* should read "0." If not, move the pointer to the "0" mark. Recheck the table height *(page 172)* if you moved the outfeed table.

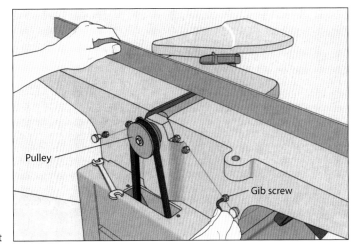

Pulley

Gib screw

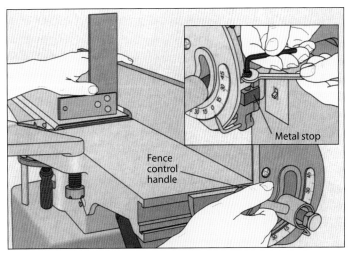

Fence control handle

Metal stop

Squaring the fence with the tables

With the fence set in its vertical position, hold a try square on the outfeed table near the cutterhead and butt the square's blade against the fence. The square should fit flush against the fence. If there is any gap between the two, slacken the fence control handle and bring the fence flush with the square *(above)*. If necessary, pivot the metal stop out of the way when making this adjustment. Then tighten the handle. The setscrew of the fence stop should be butted against the metal stop. If it is not, hold the setscrew locknut stationary with a wrench while turning the setscrew with a hex wrench *(inset)* until it touches the metal stop. Move the tilt scale indicator to "0."

Jointer Safety Tips

- Check regularly to make sure that the knives are sharp and securely fastened to the cutterhead.
- Unplug the jointer while installing knives or performing any setup operation.
- Wear appropriate safety glasses and hearing protection when operating the jointer.
- Do not joint stock with loose knots or the workpiece may catch in the cutterhead.
- Never joint stock that is less than 12 inches long.
- Do not face-joint stock that is less than ⅜ inch thick.
- Do not joint the end grain of a workpiece that is less than 6 inches wide.
- When the machine is running, keep your hands out of the area 4 inches above and to either side of the jointer's cutterhead.
- Never reach up into the dust chute unless the jointer is unplugged.

Jointer Knives

Unlike the blades of other woodworking machines, whose height and angle are adjustable, jointer knives are designed to function at just one setting: parallel to and at the same height as the machine's outfeed table. As such, the height of all the knives must be identical; a difference of as little as a fraction of an inch can compromise the jointer's ability to produce smooth, square edges.

Like all blades, jointer knives work well only when they are sharp. However, because removing a jointer knife for sharpening and then reinstalling it properly can be a time-consuming operation, many woodworkers go to great lengths to avoid changing these blades. It is possible to use an oilstone to hone the cutting edges of slightly dull knives while they are in the cutterhead. But you risk removing more metal from the cutting edges than is absolutely necessary and this can throw the knives out of alignment with the outfeed table.

There are tricks you can use to prolong the useful life of a set of knives, but once your machine begins producing uneven limp shavings or burnishing the wood, it is time to remove the knives and have them reground. Be sure to give the person doing the sharpening explicit instructions regarding the same amount of steel to be removed from each knife. Otherwise, the cutterhead may become imbalanced, causing machine vibration and also possible motor failure.

When changing your jointer knives, remove and reinstall them one at a time. Taking the blades all off at once and then installing them one after another can put stress on the cutterhead.

If you are considering replacing the knives, you can choose between high speed steel or tungsten carbide. The carbide variety offers superior performance in cutting abrasive materials such as plywood; they cost more, however. Always replace the entire set of blades, rather than individual knives. In the meantime, keep your knives clean by rubbing them occasionally with a cloth dampened in turpentine or lacquer thinner.

Changing Jointer Knives

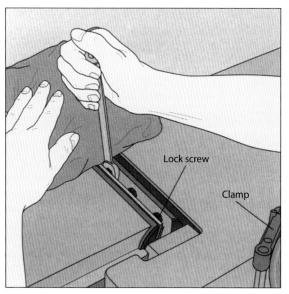

Removing an old knife
Remove the fence, then install a clamp on the rabbeting ledge to hold the guard temporarily out of the way. Use a small wood scrap to rotate the cutterhead until the lock screws securing the knife are accessible between the tables. Cover the edge of the knife with a rag to protect your hands, then use a wrench to loosen each screw in turn *(left)*. Carefully lift the knife out of the cutterhead. Remove the retaining wedge and wipe it clean.

Using a Knife-Setting Jig

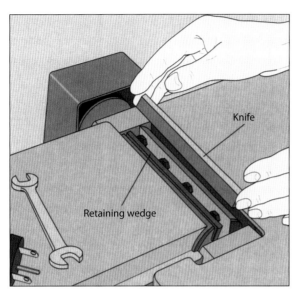

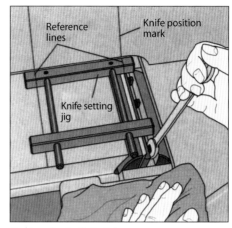

Setting the knife height
Remove an old knife and install a new one *(page 174)*. Use a small wedge to rotate the cutterhead until the edge of the new knife is at its highest point. Then mark a line on the fence directly above the cutting edge using a square and a pencil. Position a commercial knife-setting jig on the outfeed table, aligning the reference line on the jig arm with the marked line on the fence, as shown. Mark another line on the fence directly above the second reference line on the jig arm. Remove the jig and extend this line across the outfeed table. Reposition the jig on the table, aligning its reference lines with the marked lines on the fence; the jig's magnetic arms will hold the knife at the correct height while you use a wrench to tighten the lockscrews.

Installing a new knife
Insert the retaining wedge in the cutterhead, centering it in the slot with its grooved edge facing up; make sure that the heads of the lock screws are butted against the back edge of the slot as shown. With the beveled edge of the knife facing the outfeed table *(above)*, place it between the retaining wedge and the front edge of the slot, leaving the beveled part protruding from the cutterhead.

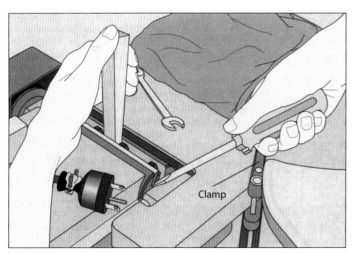

Setting the knife height
Cover the edge of the knife with a rag and partially tighten each lock screw in turn; then tighten them fully, beginning with the ones in the center and working out to the edges. Check the outfeed table height *(page 172)* in relation to the knife just installed. If the knife is set too low, loosen the lock screws slightly, then pry up the knife using a screwdriver *(above)* while holding the cutterhead stationary with a wedge; if it is too high, tap it down using a wood block. Tighten the lock screws and remove the clamp from the rabbeting ledge.

Rabbets, Chamfers, and Tapers

With a little resourcefulness, you can do more than produce square boards on a jointer. By taking full advantage of the machine's capabilities, you can shape wood with tapers and chamfers, or even cut rabbets for joinery. In fact, many woodworkers consider the jointer the best tool for cutting rabbets—at least when you are working with the grain of a workpiece.

As long as your jointer has a rabbeting ledge, it can cut rabbets along either the edge or the face of a board. Since the guard must be removed for edge rabbets on stock thicker than ¾ inch and for any rabbet along the face of a board, extra caution is essential.

Angled cuts along the corners of a workpiece, known as chamfers, are made on the jointer by tilting the fence to the required angle or with the aid of a shop-made jig. Tapers are also straightforward. With a stop block clamped to each table, you can cut stopped tapers that leave square ends for joining to a tabletop or seat, or for carving into a decorative foot.

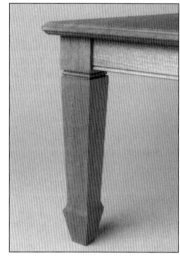

A leg tapered on the jointer provides graceful support for this table.

Rabbeting on the Jointer

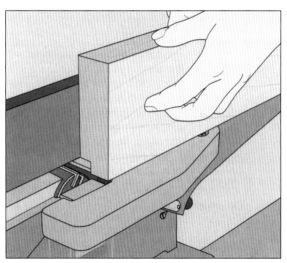

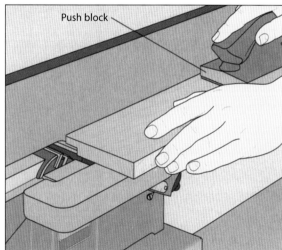

Push block

Cutting rabbets

Mark cutting lines for the width and depth of the rabbet on the leading end of the workpiece. Align the width mark with the ends of the knives, then position the fence flush against the workpiece. Set the cutting depth no deeper than ¼ inch. For a rabbet along a board edge *(above, left)*, feed the workpiece from above with your right hand while your left hand maintains pressure against the fence. Increase the cutting depth by increments no deeper than ¼ inch and make additional passes if necessary.

For a rabbet along a board face *(above, right)*, guide the workpiece near its front end with your left hand, while using a push block to apply downward pressure and keep the workpiece flat on the tables. Slowly feed the workpiece across the knives, then deepen the rabbet, if necessary.

Woodworking MACHINES

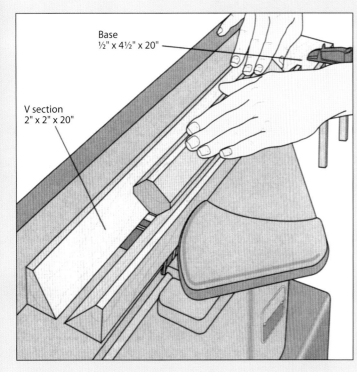

Base
½" x 4½" x 20"

V section
2" x 2" x 20"

A V-Block Jig

To cut a series of chamfers on the jointer, use this simple shop-made jig. Refer to the illustration shown at left for suggested dimensions.

Begin the V section of the jig by bevel cutting 2-by-2s. Position the two cut pieces so that they extend beyond one end of the base by about 6 inches, and have a ½-inch gap between them. Attach the two pieces through the base with countersunk screws to avoid scratching the jointer table when the jig is clamped in place.

To use the jig, clamp it in place with one end of the base aligned with the cutterhead-end of the infeed table. Lower the infeed table to the maximum depth of cut, typically ½ inch. Seat the workpiece in the gap of the jig, then feed it across the knives with your right hand, while holding it firmly in the V with your left hand.

Making a Simple Taper

Setting up and starting the cut

Use a marking gauge to outline the taper on the workpiece *(inset)*; then mark lines on the four faces of the stock to indicate where the taper will begin. Install a clamp on the rabbeting ledge to hold the guard out of the way. Set a ⅛-inch depth of cut and, holding the workpiece against the fence, align the taper start line with the front of the outfeed table. Butt a stop block against the other end of the workpiece and clamp it to the infeed table. To start each pass, carefully lower the workpiece onto the knives while holding the workpiece firmly against the fence and making sure that your hands are on the infeed side of the knives *(right)*. Straddle the fence with your right hand, using your thumb to keep the workpiece flush against the stop block.

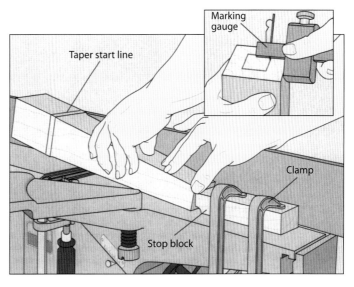

Marking gauge

Taper start line

Clamp

Stop block

Cutting the taper

Use a push stick to feed the workpiece across the cutterhead. With your right hand, apply downward pressure on the trailing end of the workpiece; use your left hand to keep the workpiece flush against the fence *(right)*. Make as many passes across the knives as necessary to complete the taper on the first face of the workpiece. To cut the remaining faces, rotate the workpiece clockwise 90° and make repeated passes over the cutterhead until you have trimmed the stock down to the taper marks.

Jointing a Stopped Taper

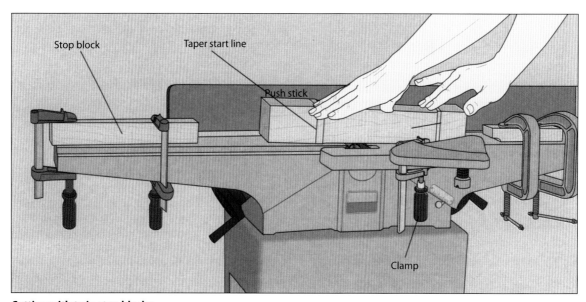

Cutting with twin stop blocks

Mark lines on all faces of the workpiece to indicate where the tapering will begin and end. Install a clamp on the rabbeting ledge to hold the guard out of the way. Set a ⅛-inch depth of cut, then butt the workpiece against the fence with the taper start line ¾ inch behind the front of the outfeed table. (The extra ¾ inch will compensate for the fact that, when the infeed table is lowered later, it will also slide back slightly.) Butt a stop block against the end of the workpiece and clamp it to the infeed table. Next align the taper end line with the back end of the infeed table. Butt a second stop block against the other end of the workpiece and clamp in place. To make the first pass, lower the workpiece onto the knives, keeping it flush against the fence and the stop block on the infeed table. Feed the workpiece using the thumb of your right hand *(above)*, fingers straddling the fence; use your left hand to press the workpiece against the fence and down on the knives. Keep both hands well above the cutterhead. Make one pass on each face, then lower the infeed table ⅛ inch and repeat the process on all four sides. Continue, increasing the cutting depth until the taper is completed.

Planer

For smoothing rough stock, planing a glued-up panel or reducing the thickness of a board uniformly, the planer is the ideal woodworking machine. Its main function is to plane wood from a board, producing a smooth surface that is parallel with the opposite face.

Planers are easy to use, but keep the following points in mind to get the best results. Always feed stock into the knives following the direction of grain. Although the maximum depth of cut for most planers is 1/8 inch, limit each pass to 1/16 inch and make multiple passes.

Some of the tasks you perform on the jointer cannot be duplicated on the planer. You cannot, for example, straighten out a warped board. Since the planer produces parallel surfaces, warped stock will emerge thinner from the machine, but just as warped.

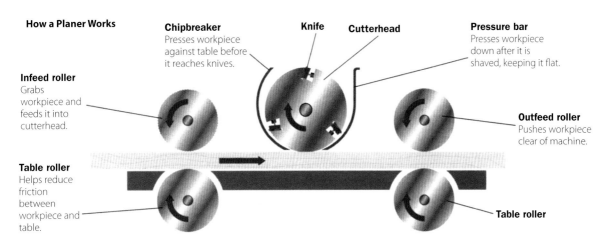

How a Planer Works

Chipbreaker Presses workpiece against table before it reaches knives.

Knife

Cutterhead

Pressure bar Presses workpiece down after it is shaved, keeping it flat.

Infeed roller Grabs workpiece and feeds it into cutterhead.

Outfeed roller Pushes workpiece clear of machine.

Table roller Helps reduce friction between workpiece and table.

Table roller

Planing a Board

Using the planer

To set the cutting depth, lay the workpiece on the table and align its end with the depth guide. For a typical 1/16-inch depth of cut, turn the table adjustment handle until the top of the board just clears the bottom of the guide *(inset)*. To make a pass through the planer, stand to one side of the workpiece and use both hands to feed it slowly into the infeed roller, keeping its edges parallel to the table edges. Once the infeed roller grips the workpiece and begins pulling it past the cutterhead, support the trailing end of the stock to keep it flat on the table *(left)*. As the trailing end of the workpiece reaches the planer's table, move to the outfeed side of the machine. Support the workpiece with both hands until it clears the outfeed roller. To prevent stock from warping, plane from both sides of a workpiece rather than removing thickness from one side only.

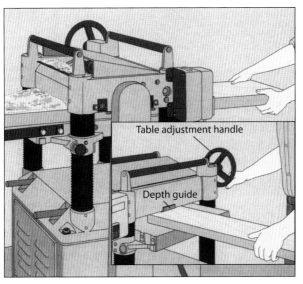

Table adjustment handle

Depth guide

Planers

Cleaning planer rollers

Planer feed rollers can get dirty quickly when planing pitch-filled softwoods such as pine. Periodically use mineral spirits or a solution of ammonia and water with a brass-bristled brush to clean metal feed rollers of pitch and resin. Clean rubber feed rollers with a sharp cabinet scraper *(right)*.

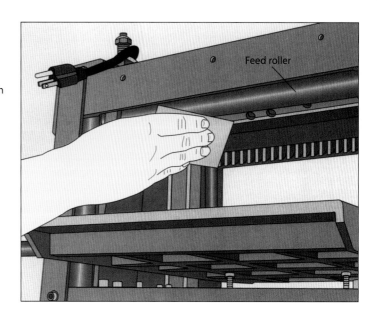

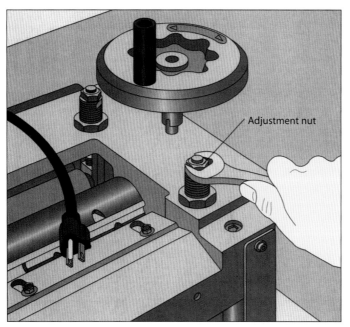

Adjusting feed rollers

Sometimes it is necessary to increase pressure on a planer's feed rollers, as when planing narrow stock or when stock slips as it is fed into the machine. In either case, the infeed roller should firmly grip the board. (Some planers feature a serrated metal infeed roller; in this case the pressure should be enough to move the board but not so much that the rollers leave a serrated pattern in the board after it is planed.) On most planers, the feed rollers are adjusted by turning spring-loaded screws on top of the machine. For the model shown, remove the plastic caps and adjust the hex nuts with an open-end wrench *(left)*. Make sure after adjusting the feed rollers that the table is parallel to the rollers. If the rollers do not carry the wood smoothly through the planer after adjustments, clean the rollers or wax the table.

Woodworking Machines

Checking the table for level
To check if your planer's table is level and parallel to the cutterhead, run two jointed strips of wood of the same thickness through opposite sides of the machine *(left)*, then compare the resulting thicknesses. If there is a measurable difference, adjust the table according to the manufacturer's instructions. If your model of planer has no such adjustment, reset the knives in the cutterhead so that they are slightly lower at the lower end of the table to compensate for the error.

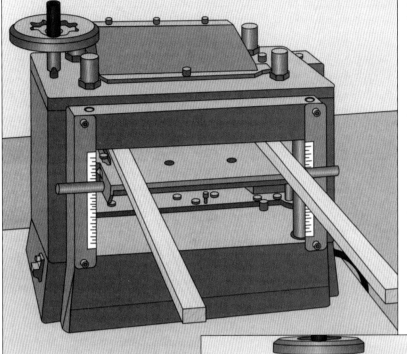

Lubricating the height adjustment
To ensure smooth operation, periodically clean the planer's height adjustment mechanism, first using a clean, dry cloth to remove sawdust and grease. Then lubricate the threads with a Teflon™-based lubricant or automotive bearing grease; oil should be avoided as it may stain the wood.

Lathes and Shapers

The drive centers of a lathe should be kept as sharp as your turning tools. If the spurs or point of a drive center are dull or chipped, they will not grip the workpiece properly. Drive centers are sharpened on a bench grinder or with a file *(left)*. A 35° bevel on the underside of each spur works best.

Lathes

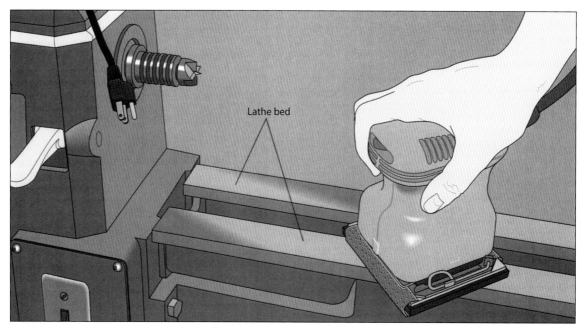

Lathe bed

Sanding the lathe bed
If your shop is in a humid climate, the bed of your lathe may develop a thin layer of rust which can prevent the tailstock and tool rest from sliding smoothly. To keep the lathe bed clean, remove any rust as soon as it appears by sanding the bed with fine sandpaper *(above)*, 200-grit or finer, then apply a paste wax.

Woodworking Machines

Draw-filing the tool rest

Because it is made of softer steel than the turning tools used against it, the bearing surface of the tool rest will develop low spots, marks, and nicks with constant use. If not remedied, these imperfections will be transferred to the workpieces you turn, or cause the tool to skip. You can redress a tool rest easily with a single-cut bastard mill file. Draw-file the rest by holding the file at an angle and pushing it across the work from right to left in overlapping strokes *(right)*. Continue until you have removed the nicks and hollows, then smooth the surface with 200-grit sandpaper or emery cloth.

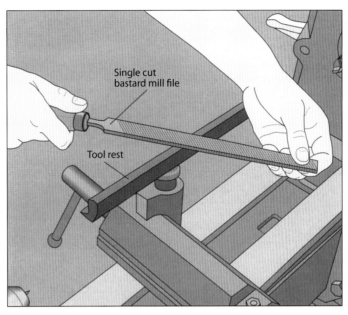

Checking for center alignment

Turning between centers requires precise alignment of drive centers between head-stock and tailstock, otherwise you will produce off-center turnings. To see if the drive centers line up, insert a four-spur drive center in the headstock and a live center in the tailstock. Slide the tailstock along the bed up to the headstock *(left)*. The points of the drive centers should meet exactly. If they do not, you may have to shim the tailstock or file down its base.

Shapers

Checking for spindle runout

Set a magnetic-base dial indicator face up on the shaper table so the plunger of the device contacts the spindle. Calibrate the gauge to zero following the manufacturer's instructions. Then turn the spindle slowly by hand *(right)*. The dial indicator will register spindle runout—the amount of wobble that the spindle will transfer to the cutter. Perform the test at intervals along the length of the spindle, adjusting its height ½ inch at a time. If the runout exceeds 0.005 inch for any of the tests, replace the spindle.

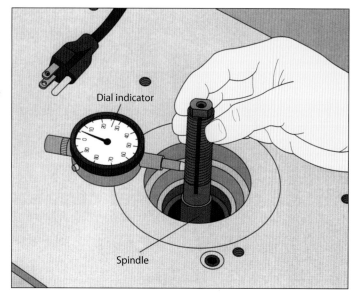

Squaring the fences

The two halves of a shaper fence—or a router table fence—must be perfectly parallel, otherwise your cuts will be uneven. To square the fences on a shaper, first loosen the fence locking handles. Hold a straightedge against the fences. The two halves should butt against the straightedge *(left)*. If not, add wood shims behind the fences until they are parallel.

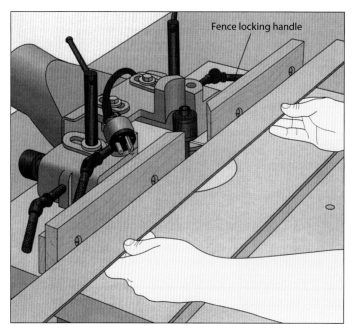

Woodworking Machines

Other Tools

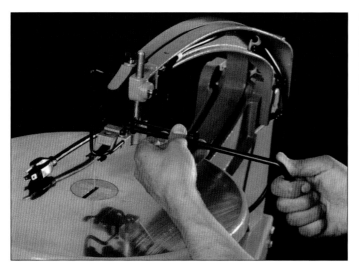

Because a scroll saw blade is held in clamps that pivot on the end of the saw's arms during a cut, replacing a blade is a tricky task that risks stretching and snapping the delicate cutting edge. The model of scroll saw shown at left features a unique blade-changing wrench that holds the blade clamps steady as the blade is tightened.

Scroll Saws

Checking blade tension

The blades of a scroll saw—like those of a band saw—require proper tension to cut effectively. Too little tension will cause excessive vibration and allow the blade to wander during the cut. Too much tension can lead to blade breakage. To adjust blade tension on the model shown, first tilt the blade tension lever forward. Then adjust the blade tension knob *(right)* to increase or decrease blade tension. Tilt the blade tension lever back and test the blade. It should deflect about ⅛ inch when pushed from side to side. Pluck the blade and remember the sound. It will allow you to tension the blade quickly in future. Always adjust the tension when you change blades.

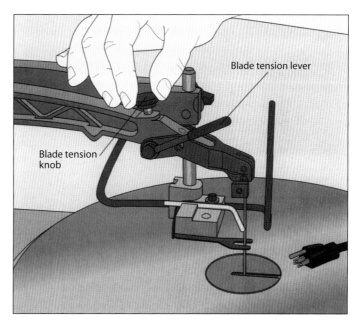

Squaring the blade

To square the scroll saw's blade to the table, butt a combination square against the blade as shown. The square should fit flush against the blade. If there is a gap, loosen the table lock knob and adjust the nut on the 90° stop until the table is level and there is no gap between the square and the blade. Tighten the lock knob.

Shop Tip

Installing an air pump

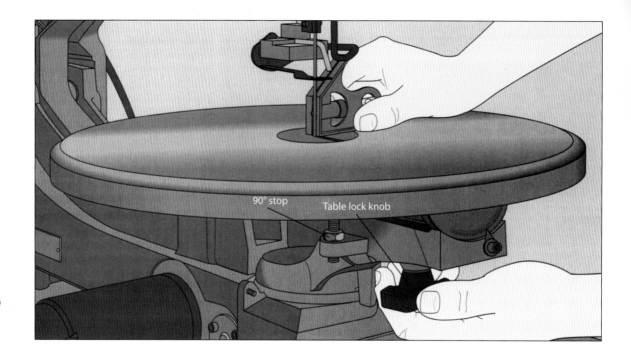

Older scroll saws and some foreign models often come without a sawdust blower, a device that keeps the cutting line clear while the saw is in use. A simple electric aquarium pump and some copper tubing (right) can do the trick at a fraction of the cost of a sawdust blowing attachment. Simply insert a 10- to 12-inch length of copper tubing into the pump's plastic air hose, making an airtight seal. Tape the hose to the saw's upper arm, and bend the copper end to point at the blade. Pinch the end of the tube slightly to direct the air and increase its pressure.

Belt-and-Disc Sanders

Testing for trueness
To measure whether the wheel is true, first remove any abrasive discs. Connect a dial indicator to a magnetic base and set the base on the tool's disc table. Place the device so that its arm contacts the disc and turn on the magnet. Calibrate the dial indicator to zero following the manufacturer's instructions. Turn the sanding disc by hand, and read the result *(right)*. The dial indicator will register the trueness of the wheel. Perform the test at various points around the disc. If the amount of wobble exceeds 0.005 inch for any of the tests, adjust the motor position or have the bearings replaced.

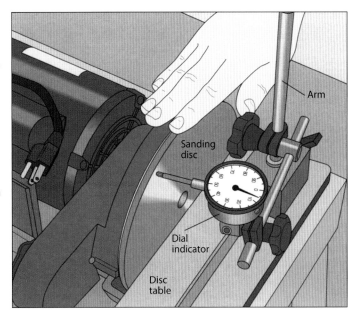

Tracking the sanding belt
To straighten a sanding belt that is not tracking true, turn the belt-and-disc sander's tracking knob clockwise or counterclockwise while the tool is running *(left)*. To correct severe tracking problems, unplug the tool and release tension on the sanding belt by pushing down on the tracking knob. Center the belt on the pulleys and release the knob. Then turn on the tool and adjust the tracking knob as required. Always track the belt when changing belts or installing a new one.

Index

angle cuts, 83, 116, 137–39
angled holes, 159, 161

band saw blades, 32, 44–48, 105–6
　　about, 105
　　changing, 106
　　folding/storing, 48
　　prolonging life of, 30
　　removing/installing, 106
　　repairing, 46–47
　　sharpening, 44–45
　　sizes and turning radiuses, 105
　　squaring table and, 103
　　tensioning and tracking, 101
　　types and functions, 105
band saw cuts
　　angle cuts, 116
　　circle-cutting jig, 111
　　crosscutting, 115
　　curves, 94, 95, 107–11
　　on cylinders, 113, 115, 118
　　dovetail joints, 119–21
　　on duplicate pieces, 118
　　joints, 119–21
　　multiple curves, 108
　　pattern sawing, 109
　　resawing, 114
　　ripping, 112–13
　　rounding corners, 110
　　shop-made rip fence for, 113
　　stack sawing, 118
　　straight cuts, 112–15
　　taper cuts, 117
　　using miter gauge, 115
　　using stop block, 118
band saws, 94–121
　　about: overview of, 94
　　adjusting guide assemblies, 102
　　adjusting table stop, 103
　　aligning wheels, 100
　　anatomy of, 96–97
　　checking wheels, 98–99
　　Dave Sawyer on, 10
　　features of, 94
　　installing heat-resistant guide blocks, 30, 45, 102
　　maintaining, 12–15
　　out-of-round wheels, 99
　　safety tips, 104
　　setting guide blocks, 102
　　setting thrust bearings, 102
　　setting up, 98–103
　　sizes and depth of cuts, 94, 112
　　squaring table, 103
　　wheel bearings, 98
belt-and-disc sanders, 187
bevel cuts. See angle cuts
bits
　　gallery of, 32
　　tools/accessories for sharpening, 33
blades. See also sharpening blades and bits; specific tools
　　gallery of, 32
　　tools/accessories for sharpening, 33
book overview, 6–8
box joints, 91
brush assembly, replacing, 26

circular saws
　　anatomy of, 25
　　blades, 32, 41–43
　　blade-setting jig, 33
　　blade-sharpening jig, 33
　　changing blades, 42
　　cleaning blades, 42
　　keeping blade perpendicular, 16, 17
　　maintaining, 16–17
　　sharpening blades, 43
　　solvent for cleaning blades, 32, 42
cleaning solvent for blades and bits, 32, 42
cord, replacing, 28–29
crosscutting
　　with band saw, 115
　　checking for square, 78
　　jig for, 81
　　with radial arm saw, 136
　　repeat cuts on table saw, 79–80
　　with table saw, 74–82
　　wide panels, 82
curves, cutting, 94, 95, 107–11

dado cuts
　　adding blades and chippers, 142
　　grooves, 84, 85–86, 144, 146
　　head types, 84
　　installing heads, 85, 142
　　rabbets, 84, 86, 144
　　on radial arm saw, 142–47
　　repeat cuts, 143
　　stopped grooves, 84, 87, 147
　　on table saw, 84–87
　　types of, illustrated, 84
dovetail joints, 119–21
dowels and dowel cutter, 163
drill bits, 37–40, 154–56
　　brad-point, 39–40, 154
　　Forstner, 38, 154
　　gallery of, 32
　　hole saws and, 155
　　multi-spur, 39
　　plug cutter and, 155
　　sharpening, 37–40
　　sharpening jig for, 33, 37
　　spade, 40, 155
　　twist, 30, 31, 37, 154
drill presses, 148–67
　　about: overview of, 148
　　aligning table, 152, 156, 159
　　anatomy of, 150–51
　　angled holes with, 159
　　bits and accessories, 154–56. See also drill bits
　　boring into cylinders with, 161
　　changing belt position, 152
　　clamping work to, 156
　　column-mounted accessory rack, 155
　　compound angles with, 161
　　correcting chuck runout, 153
　　deep holes with, 160
　　dowels, plugs, tenons with, 163
　　features of, 148
　　installing chisel and bit, 164
　　Judith Ames on, 9
　　maintaining, 12–15
　　mortising techniques, 164–66
　　pocket hole jig for, 162
　　removing/installing bits, 154
　　replacing chuck, 153
　　sanding with, 148, 167
　　setting belt tension, 152
　　setting up and safety, 152–53
　　sizes and depth of cuts, 148
　　squaring chisel, 165
　　squaring table, 152, 156
　　stopped holes with, 157
　　straight holes with, 157–58
　　tilting table jig for, 159
　　uniformly spaced holes with, 158
drive belts, maintaining, 14

electric drills
　　anatomy of, 19
　　bits for. See drill bits

grooves, 84, 85–86, 144, 146. See also stopped grooves

hand tools. See maintaining hand tools; specific tools
hole saws, 155
honing guide, 32

jointer/planer knives, 49–55, 174–75
　　changing, 174–75
　　honing guide, 32
　　honing jointer knives, 49–50
　　installing jointer knives, 54, 55
　　removing/installing, 174–75
　　setting jigs, 33
　　sharpening jig, 33
　　sharpening jointer knives, 51–55
　　sharpening planer knives, 55
jointers, 168–78
　　about: overview of, 168
　　adjusting tables, 172–73
　　anatomy of, 170–71
　　changing knives, 174–75
　　cutting tapers with, 168, 169
　　features of, 168
　　jointing stopped taper (with twin stop blocks), 178
　　knives, 32, 49–55
　　maintaining, 12–15
　　Mark Duginske on, 11
　　rabbets on, 176
　　setting up and safety, 172–73
　　squaring fence with tables, 173
　　table height, 172
　　tapers on, 177–78

188

V-block jig for, 177
joints
    box, 91
    dovetail, 119–21
    lap, 93
    mortise-and-tenon, 56, 57, 92

lap joints, 93
lathes and shapers, 182–84

maintaining hand tools, 16–29
    about: overview of, 16
    motor and power components, 16, 26–29
    tips and schedules, 18
maintaining stationary tools, 12–15, 180–81. See also specific tools
miter cuts. See angle cuts
molding head and knives, 32, 36, 88–90
moldings, making, 88–90
mortise-and-tenon joints, 56, 57, 92
mortising techniques (drill press), 164–66
motor and power components
    replacing brush assembly, 26
    replacing cord, 28–29
    replacing switch, 27

planer knives
    honing guide, 32
    sharpening, 55
    sharpening jig, 33
planers, 179–81. See also jointer/planer knives
    adjusting and maintaining, 180–81
    features of, 179
    how they work, 179
    using (planing boards), 179
plate joiners, anatomy of, 24
plug cutter, 155
plugs, 163
pocket hole jig, 162
power cord, replacing, 28–29

rabbets, 84, 86, 144, 176
radial arm saw blades, 132–33
    anatomy of, 132
    carbide-tipped, 132
    changing, 133
    cleaning, 132
    correcting heel of, 130–31
    removing/installing, 133
    squaring with table, 129
radial arm saw cuts
    angle cuts (miter, bevel, compound), 122, 123, 137–39
    crosscutting, 136
    dado cuts, 142–47
    hold-down device for, 140
    large panels, 141
    out-rip configuration, 141
    repeat cuts, 136
    ripping, 134, 140–41
    safety tips, 134–35
radial arm saws, 122–47
    about: overview of, 122
    accessories for, 132
    adjusting clamps, 126–28
    adjusting table, 126

anatomy of, 124–25
anti-kickback devices, 135
auxiliary fence for, 145
auxiliary table for, 131, 145
carriage roller bearings, 128
column-to-base tension, 129
features of, 122
hold-down device for, 135, 140
installing fence, 131
maintaining sliding mechanism, 128–29
safety tips, 134–35
setting arm perpendicular to fence, 130
setting up, 126–31
sharpening molding knives, 36
specialty guard, 135
squaring blade with table, 129
resawing thick stock, 76, 114
resin solvent, 32, 42
ripping, 72–75
    with band saw, 112–13
    defined, 72
    jig for repeat narrow cuts, 75
    large panels, 74, 141
    narrow strips, 75
    with radial arm saw, 134, 140–41
    with table saw, 72–75
routers
    anatomy of, 20
    bit sharpener, 33
    bits and shaper cutters, 32, 34–35
    checking collet for runout, 21
    correcting out-of-round sub-base, 16
    shaper cutters, 32, 35
    sharpening bits and cutters, 34–35
    storage rack for shaper cutters, 35
    using dial indicator and magnetic base with, 21
    using feeler gauge with, 21

saber saws
    anatomy of, 22
    extending blade life, 23
    squaring blade, 23
sanders, belt-and-disc, 187
sanding, with drill press, 167
scroll saws, 185–86
shaper cutters, 32, 35
shapers, 184
sharpening blades and bits, 30–55
    about: overview of, 30
    band saw blades, 44–45
    circular blades, 41–43
    drill bits, 37–40
    gallery of blades and bits, 32
    jointer and planer knives, 49–55
    molding knives, 36
    router bits and shaper cutters, 34–35
    solvent for cleaning before, 32, 42
    tools and accessories for, 33
    twist bits, 30, 31
soldering band saw blades, 46–47
stationary tools, maintaining, 12–15
stopped grooves, 84, 87, 147
switches
    hands-free "Off" switch, 80
    maintaining, 15

replacing, 27
table saw blades
    adjusting height and tilt, 65, 71
    aligning table and, 61–62
    blade guard assembly, 67
    carbide-tipped, 68–69
    changing, 70
    cleaning, 42
    installing/removing, 41, 70
    sharpening, 43
    types and functions, 68–69
table saw cuts
    angle cuts, 83
    box joints, 91
    crosscutting, 78–82
    dado cuts (and making dadoes), 84–87
    grooves, 84, 85–86
    lap joints, 93
    large panels, 74, 82
    moldings, 88–90
    mortise-and-tenon joints, 56, 57, 92
    rabbets, 84, 86
    repeat cuts using miter gauge, 80
    repeat cuts using rip fence as guide, 79
    resawing thick stock, 76
    ripping, 72–75
    safety tips, 66–67
    stopped grooves, 84, 87
    taper cuts, 77
    using miter gauge, 80, 83
table saws, 56–93
    about: overview of, 56, 60
    aligning rip fence, 64
    aligning/squaring miter gauge, 63
    anatomy of, 58–59
    blade guard assembly, 67
    features of, 56
    Giles Miller on, 8
    hands-free "Off" switch, 80
    leveling table insert, 65
    maintaining, 12–15
    miter extension, 56
    models and sizes, 56
    safety tips, 66–67
    setting up, 60
    sharpening molding knives, 36
    testing for square, 64
tabletops, maintaining, 15
taper cuts, 77, 117, 177–78
tenons, 163. See also mortise-and-tenon joints

V-block jig, 177

THE *Missing* SHOP MANUAL SERIES

These are the manuals that should have come with your new woodworking tools. In addition to explaining the basics of safety and set-up, each *Missing Shop Manual* covers everything your new tool was designed to do on its own and with the help of jigs & fixtures. No fluff, just straight tool information at your fingertips.

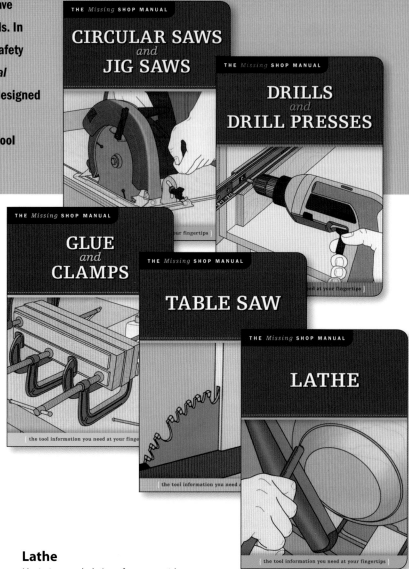

Circular Saws and Jig Saws
From ripping wood to circle cutting, you'll discover the techniques to maximize your saw's performance.

ISBN 978-1-56523-469-7
$9.95 USD • 88 Pages

Drills and Drill Presses
Exert tips and techniques on everything from drilling basic holes and driving screws to joinery and mortising.

ISBN 978-1-56523-472-7
$9.95 USD • 104 Pages

Glue and Clamps
Learn how to get the most out of your clamps and that bottle of glue when you're carving, drilling, and building furniture.

ISBN 978-1-56523-468-0
$9.95 USD • 104 Pages

Table Saw
Whether you're using a bench top, contractor or cabinet saw, get tips on everything from cutting dados and molding to creating jigs.

ISBN 978-1-56523-471-0
$12.95 USD • 144 Pages

Lathe
Maximize your lathe's performance with techniques for everything from sharpening your tools to faceplate, bowl, and spindle turning.

ISBN 978-1-56523-470-3
$12.95 USD • 152 Pages

BUILT to LAST TIMELESS WOODWORKING PROJECTS

Discover the timeless woodworking projects that are *Built to Last*. These are the classic and enduring woodworking projects that stand the test of time in form and function and in the techniques they employ. Ideal for all skill levels, the *Built to Last* series represents the pieces and projects that every woodworker should build in a lifetime.

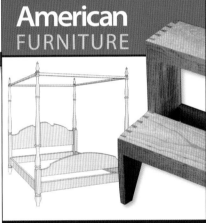

Shaker Furniture
Delve into this old-world style with 12 enduring projects. Includes step-by-step instruction and crisp drawings of each project.
ISBN 978-1-56523-467-3
$19.95 USD • 144 Pages

American Furniture
These classic furniture projects are the reason many people get into woodworking. Build your skills and confidence as you build beautiful furniture.
ISBN 978-1-56523-501-4
$19.95 USD • 136 Pages

Outdoor Furniture
12 practical projects to furnish your outdoor living space. Organized, step-by-step instruction will help you build smart and get it done right.
ISBN 978-1-56523-500-7
$19.95USD • 144 Pages

Look for These Books at Your Local Bookstore or Woodworking Retailer
To order direct, call **800-457-9112** or visit *www.FoxChapelPublishing.com*

By mail, please send check or money order + $4.00 per book for S&H to:
Fox Chapel Publishing, 1970 Broad Street, East Petersburg, PA 17520

Back to Basics
Straight Talk for Today's Woodworker

Get *Back to Basics* with the core information you need to succeed. This new series offers a clear road map of fundamental woodworking knowledge on sixteen essential topics. It explains what's important to know now and what can be left for later. Best of all, it's presented in the plain-spoken language you'd hear from a trusted friend or relative. The world's already complicated—your woodworking information shouldn't be.

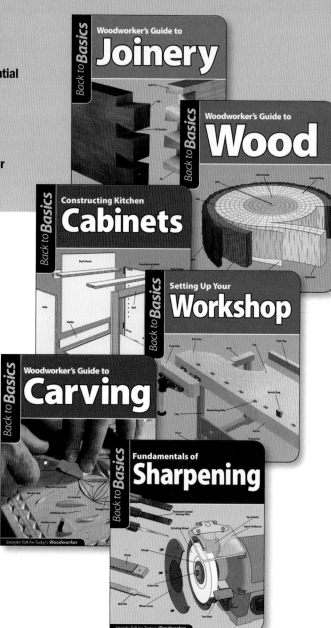

Woodworker's Guide to Joinery
ISBN 978-1-56523-462-8
$19.95 USD • 200 Pages

Woodworker's Guide to Wood
ISBN 978-1-56523-464-2
$19.95 USD • 160 Pages

Constructing Kitchen Cabinets
ISBN 978-1-56523-466-6
$19.95 USD • 144 Pages

Setting Up Your Workshop
ISBN 978-1-56523-463-5
$19.95 USD • 152 Pages

Woodworker's Guide to Carving
ISBN 978-1-56523-497-0
$19.95 USD • 160 Pages

Fundamentals of Sharpening
ISBN 978-1-56523-496-3
$19.95 USD • 128 Pages

Look for These Books at Your Local Bookstore or Woodworking Retailer
To order direct, call **800-457-9112** or visit *www.FoxChapelPublishing.com*

By mail, please send check or money order + $4.00 per book for S&H to:
Fox Chapel Publishing, 1970 Broad Street, East Petersburg, PA 17520